AF581714

Mediterranean Diet and Physical Activity as Healthy Lifestyles for Human Health

Mediterranean Diet and Physical Activity as Healthy Lifestyles for Human Health

Editor

Daniela Bonofiglio

MDPI • Basel • Beijing • Wuhan • Barcelona • Belgrade • Manchester • Tokyo • Cluj • Tianjin

Editor
Daniela Bonofiglio
University of Calabria
Italy

Editorial Office
MDPI
St. Alban-Anlage 66
4052 Basel, Switzerland

This is a reprint of articles from the Special Issue published online in the open access journal *Nutrients* (ISSN 2072-6643) (available at: https://www.mdpi.com/journal/nutrients/special_issues/Mediterranean_Diet_Human_Health).

For citation purposes, cite each article independently as indicated on the article page online and as indicated below:

LastName, A.A.; LastName, B.B.; LastName, C.C. Article Title. *Journal Name* **Year**, *Volume Number*, Page Range.

ISBN 978-3-0365-5095-4 (Hbk)
ISBN 978-3-0365-5096-1 (PDF)

Cover image courtesy of Daniela Bonofiglio

Contents

About the Editor

Daniela Bonofiglio

Daniela Bonofiglio, MD endocrinologist, is currently a Full Professor of Biotechnology and Methods in Laboratory Medicine at the Department of Pharmacy, Health and Nutritional Sciences, University of Calabria, Italy. Her scientific interests are devoted to dissecting the molecular mechanisms involved in the initiation and progression of endocrine-related cancers with the main goal of identifying novel markers and potential therapeutic targets for these diseases. Her achievements contributed to advancing scientific knowledge on peroxisome proliferator-activated receptor gamma (PPARγ) biology, focusing on the inhibiting role of natural and synthetic ligands of PPARγ in cancer growth and progression. In the context of the natural products field, her research activity has also been focused on the beneficial role of different food compounds on human health. Specifically, in the last two decades, she has investigated the iodine status of young and adult populations living in sufficient and deficient iodine areas and the impact of adherence to the Mediterranean diet on the biochemical and inflammatory parameters in adolescents and adults from Southern Italy. She is the author of more than 100 peer-reviewed papers and 6 book chapters related to the subjects of endocrinology, metabolism, and cancer.

Link to the publication:

Orcid (http://orcid.org/0000-0002-4142-0496)

Scopus (https://www.scopus.com/authid/detail.uri?authorId=6603246592)

PubMed (https://www.ncbi.nlm.nih.gov/pubmed/?term=bonofiglio+d)

Editorial

Mediterranean Diet and Physical Activity as Healthy Lifestyles for Human Health

Daniela Bonofiglio [1,2]

[1] Department of Pharmacy, Health and Nutritional Sciences, University of Calabria, 87036 Rende, Cosenza, Italy; daniela.bonofiglio@unical.it; Tel.: +39-0984-496208; Fax: +39-0984-496203
[2] Centro Sanitario, University of Calabria, 87036 Rende, Cosenza, Italy

Health status is influenced by several factors, such as proper dietary pattern and regular physical activity (PA), which are crucial elements of lifestyle in terms of the prevention and treatment of metabolic and chronic diseases in all stages of life and particularly during childhood and adolescence.

In the last decades, cultural globalization and urbanisation have led to the "Westernization" of lifestyles, characterized by the increased consumption of foods with high quantities of refined carbohydrates, sugars, salt, saturated fats, proteins, as well as poor fruits and vegetables, and by increasing sedentariness. This phenomenon has generated a "nutritional transition" whereby obesity and diet-related chronic diseases represent new challenges for public-health systems. In contrast to this global trend and in the context of healthy eating habits, the dietary pattern inspired by Mediterranean Diet (MD) principles is associated with multiple health benefits, due to its protective effects against a wide range of chronic and metabolic diseases, including obesity, diabetes mellitus, cardiovascular and neurodegeneretive diseases, and cancers [1–3]. The traditional MD, usually consumed among the populations bordering the Mediterranean Sea many years ago, has entered the medical literature following publications of Seven Countries Study's results by the legendary Ancel Keys [4]. The MD pattern is characterized by high intakes of vegetables, legumes, fruits and nuts, cereals (that in the past were largely unrefined), and dairy products; high consumption of extra virgin olive oil; low intakes of saturated lipids; moderately high intakes of fish and poultry; low intakes of meat and sweets; and regular but moderate intake of wine generally during meals [5]. Seasonality, biodiversity, traditional and local food products are also important components in this eating model. In addition, the MD has qualitative cultural and lifestyle elements, such as conviviality, culinary activities, adequate rest, and physical activity [5]. Regarding PA, the WHO (World Health Organisation) recommends performing moderate-intensity levels of PA for $\geq$150 min/week, and vigorous-intensity levels of PA for $\geq$2 days/week to have health benefits [6]. Overall, the MD, including PA as an integral part of the traditional Mediterranean lifestyle, is also a universal, cultural, social and spatial heritage of all civilizations living around the Mediterranean basin registered by the UNESCO (United Nations Educational, Scientific and Cultural Organization) as an immaterial human heritage in 2010. Despite its increasing popularity worldwide, adherence to the MD model is decreasing due to multifactorial influences determining the ongoing erosion of this dietary pattern and cultural heritage worldwide, even in the Mediterranean area. Thus, the need to investigate the current dietary habits and to increase population awareness of the importance of MD pattern is becoming urgent.

The Special Issue of Nutrients entitled "Mediterranean Diet and Physical Activity as Healthy Lifestyles for Human Health" was devoted to collecting original research and reviews of literature concerning the adherence to the MD and various health outcomes. New information has been added in this field by means of ten articles, with nine original papers and one narrative review.

Citation: Bonofiglio, D. Mediterranean Diet and Physical Activity as Healthy Lifestyles for Human Health. *Nutrients* **2022**, *14*, 2514. https://doi.org/10.3390/nu14122514

Received: 16 May 2022
Accepted: 8 June 2022
Published: 17 June 2022

A widely considered aspect was the evaluation of the MD adherence in association with PA in different target populations, such as children, adolescents, and adults, as well as populations from different geographical location worldwide.

Firstly, Ceraudo et al. [7], evaluating the impact of the Mediterranean food choices on serum metabolic profile in healthy adolescents performing different intensity levels of PA, showed that active subjects who consume typical Mediterranean foods had a better serum metabolic profile, suggesting that adhering to the MD pattern and performing PA led to a significant reduction in glucose and lipid profile, thus making the MD and PA a winning combination for health status.

A comparative study carried out among 2722 individuals aged 2 to 24 years living in Croatia showed low adherence to the MD over the entire sample [8]. Specifically, Matana et al., found that the prevalence rate of poor adherence to the MD increased with higher education stage, while the lowest rate was observed for the children enrolled in kindergartens [8]. In agreement with other studies conducted in the same age range [9,10], individuals physically active were also higher adherent to the MD, fostering the positive association of both factors with healthy lifestyle habits.

The association between PA and adherence to the MD was also investigated in 1220 fitness-center users in Croatia [11]. Results showed that MD adherence, measured by means of the Mediterranean Diet Serving Score (MDSS) in the whole study sample (mean age was 29.1 $\pm$ 8.8 years) was 8.0, and 18.6% of participants adhered to the MD (total MDSS score $\geq$ 14). Interestingly, there was a significant positive correlation between the level of PA and the MDSS score in this population, without gender differences. Thus, from these findings it is possible to speculate that promoting adherence to the MD along with PA guidelines might provide a more comprehensive endorsement to obtain greater health benefits, over and above those acquired separately by the MD and PA.

Another Croatian group of research [12] found that MD adherence, evaluated by MDSS, had a prevalence of 28.5% in 4671 adult subjects, with higher odds in women, older subjects, and people with a higher level of objective material status. Over the study period, the absolute change in the MD score positively associated with female gender, age, higher education, and moderate physical activity, but it was negatively correlated with adherence to the MD at baseline, suggesting that these factors can be potentially targeted in order to increase MD adherence.

Among three other studies carried out in both Mediterranean [13,14] and non-Mediterranean countries [15], the first involved 1512 Spanish adults, aged 55–80 years, with overweight/obesity and metabolic syndrome. In this cross-sectional study, analyzing data from a sub-sample of the PREDIMED-Plus study, socioeconomic status was related to an unhealthy dietary pattern associated with low PA [13]. This indicates that community interventions and health policy decisions may target subsets of the population in order to promote a healthier lifestyle. The second one [14] investigated in an Italian adult sample the relationship between adherence to the MD with the self-perceived adoption of a sustainable diet, as well as demographic, socioeconomic, and behavioral variables. Results showed that MD scores were high in female subjects, in subjects with higher income and educational level, in subject who consider the MD a sustainable dietary model, as well as those who perceive themselves as following a sustainable diet. Globally, a medium adherence to the MD was found [14], in line with another recent investigation on an Italian adult population [16]. Intriguingly, eating pattern, perceived knowledge, and benefits have been evaluated by Cuoto et al. [15] in a Portuguese immigrant community living in California. Even though Portugal is geographically not in the Mediterranean basin, the MD is a settled cultural heritage of the Portuguese population. MD scores were higher in adults from the immigrant than in those from the American population, and the perceived health benefits of the diet was a key factor in adherence to the MD only in the Portuguese immigrant community. The authors reported the need for further investigations to confirm these results.

Apart from the studies evaluating adherence to the MD in different contexts, also interesting was the relationship between diet and some metabolic and chronic diseases, such as type 1 diabetes (T1D) and cancer. In this scenario, Antoniotti et al. [17] assessed the MD adherence by the Mediterranean Diet Quality Index (KIDMED) questionnaire in 65 children and adolescents with T1D in relation to metabolic control. KIDMED scores displayed average values in 58.6% of the subjects, in line with data from a non-diabetic population of Italians of the same age range [8]. Low adherence to the MD was associated with a high risk of obesity in T1D, as reported in the general pediatric and adult population [17]. Regarding the prospective relation between diet pattern and total cancer risk, Yiannakou et al. [18] examined in a prospective study the longitudinal association between the Mediterranean-style dietary pattern (MSDP) score, derived from a semi-quantitative food frequency questionnaire, and total cancer risk in 2966 participants of the Framingham Offspring Study cohort, displaying weaker association between MSDP score and cancer risk in men except in non-smokers, while higher adherence to MSDP was associated with lower cancer risk, especially among women [18], confirming that the MD as a source of bioactive compounds protects against cancer [19].

Last, the impact of geographical location on MD adherence has been also considered and discussed in a narrative review by Mattavelli et al. [20] who highlighted the relevance of geographical location and related social features in the current moderate adherence to the MD in adolescents as well as in adults, fostering awareness that will lead to the promotion of the MD as a global nutritionally balanced and healthy dietary pattern even in the countries of its origin.

Overall, all studies included in this Special Issue provide an update on the MD adherence in the population at different age stages and from different countries, highlighting some opportunities and challenges for the adoption of an MD eating pattern along with PA to reduce metabolic and chronic disease risk and to obtain greater health benefits.

Funding: This research received no external funding.

Conflicts of Interest: The author declares no conflict of interest.

References

1. Galbete, C.; Schwingshackl, L.; Schwedhelm, C.; Boeing, H.; Schulze, M.B. Evaluating Mediterranean diet and risk of chronic disease in cohort studies: An umbrella review of meta-analyses. *Eur. J. Epidemiol.* **2018**, *33*, 909–931. [CrossRef] [PubMed]
2. Dinu, M.; Pagliai, G.; Casini, A.; Sofi, F. Mediterranean diet and multiple health outcomes: An umbrella review of meta-analyses of observational studies and randomised trials. *Eur. J. Clin. Nutr.* **2018**, *72*, 30–43. [CrossRef] [PubMed]
3. Grosso, G.; Marventano, S.; Yang, J.; Micek, A.; Pajak, A.; Scalfi, L.; Galvano, F.; Kales, S.N. A comprehensive meta-analysis on evidence of Mediterranean diet and cardiovascular disease: Are individual components equal? *Crit. Rev. Food Sci. Nutr.* **2017**, *57*, 3218–3232. [CrossRef] [PubMed]
4. Montani, J.P. Ancel Keys: The legacy of a giant in physiology, nutrition, and public health. *Obes. Rev.* **2021**, *22* (Suppl. S2), e13196. [CrossRef] [PubMed]
5. Serra-Majem, L.; Tomaino, L.; Dernini, S.; Berry, E.M.; Lairon, D.; Ngo de la Cruz, J.; Bach-Faig, A.; Donini, L.M.; Medina, F.X.; Belahsen, R.; et al. Updating the Mediterranean Diet Pyramid towards Sustainability: Focus on Environmental Concerns. *Int. J. Environ. Res. Public Health* **2020**, *17*, 8758. [CrossRef] [PubMed]
6. Global Action Plan on Physical Activity 2018–2030: More Active People for a Healthier World. World Health Organization. Available online: https://apps.who.int/iris/handle/10665/272722 (accessed on 6 May 2022).
7. Ceraudo, F.; Caparello, G.; Galluccio, A.; Avolio, E.; Augimeri, G.; De Rose, D.; Vivacqua, A.; Morelli, C.; Barone, I.; Catalano, S.; et al. Impact of Mediterranean Diet Food Choices and Physical Activity on Serum Metabolic Profile in Healthy Adolescents: Findings from the DIMENU Project. *Nutrients* **2022**, *14*, 881. [CrossRef] [PubMed]
8. Matana, A.; Franić, I.; Radić Hozo, E.; Burger, A.; Boljat, P. Adherence to the Mediterranean Diet among Children and Youth in the Mediterranean Region in Croatia: A Comparative Study. *Nutrients* **2022**, *14*, 302. [CrossRef] [PubMed]
9. Arcila-Agudelo, A.M.; Ferrer-Svoboda, C.; Torres-Fernandez, T.; Farran-Codina, A. Determinants of Adherence to Healthy Eating Patterns in a Population of Children and Adolescents: Evidence on the Mediterranean Diet in the City of Mataro (Catalonia, Spain). *Nutrients* **2019**, *11*, 854. [CrossRef] [PubMed]
10. Morelli, C.; Avolio, E.; Galluccio, A.; Caparello, G.; Manes, E.; Ferraro, S.; De Rose, D.; Santoro, M.; Barone, I.; Catalano, S.; et al. Impact of Vigorous-Intensity Physical Activity on Body Composition Parameters, Lipid Profile Markers, and Irisin Levels in Adolescents: A Cross-Sectional Study. *Nutrients* **2020**, *12*, 742. [CrossRef]

11. Martinovic, D.; Tokic, D.; Martinovic, L.; Kumric, M.; Vilovic, M.; Rusic, D.; Vrdoljak, J.; Males, I.; Ticinovic Kurir, T.; Lupi-Ferandin, S.; et al. Adherence to the Mediterranean Diet and Its Association with the Level of Physical Activity in Fitness Center Users: Croatian-Based Study. *Nutrients* **2021**, *13*, 4038. [CrossRef] [PubMed]
12. Pribisalić, A.; Popović, R.; Salvatore, F.P.; Vatavuk, M.; Mašanović, M.; Hayward, C.; Polašek, O.; Kolčić, I. The Role of Socioeconomic Status in Adherence to the Mediterranean Diet and Body Mass Index Change: A Follow-Up Study in the General Population of Southern Croatia. *Nutrients* **2021**, *13*, 3802. [CrossRef] [PubMed]
13. Basora, J.; Villalobos, F.; Pallejà-Millán, M.; Babio, N.; Goday, A.; Zomeño, M.D.; Pintó, X.; Sacanella, E.; Salas-Salvadó, J. Deprivation Index and Lifestyle: Baseline Cross-Sectional Analysis of the PREDIMED-Plus Catalonia Study. *Nutrients* **2021**, *13*, 3408. [CrossRef] [PubMed]
14. Biasini, B.; Rosi, A.; Menozzi, D.; Scazzina, F. Adherence to the Mediterranean Diet in Association with Self-Perception of Diet Sustainability, Anthropometric and Sociodemographic Factors: A Cross-Sectional Study in Italian Adults. *Nutrients* **2021**, *13*, 3282. [CrossRef] [PubMed]
15. Couto, R.M.; Frugé, A.D.; Greene, M.W. Adherence to the Mediterranean Diet in a Portuguese Immigrant Community in the Central Valley of California. *Nutrients* **2021**, *13*, 1989. [CrossRef] [PubMed]
16. Caparello, G.; Galluccio, A.; Giordano, C.; Lofaro, D.; Barone, I.; Morelli, C.; Sisci, D.; Catalano, S.; Ando, S.; Bonofiglio, D. Adherence to the Mediterranean diet pattern among university staff: A cross-sectional web-based epidemiological study in Southern Italy. *Int. J. Food Sci. Nutr.* **2019**, *71*, 581–592. [CrossRef] [PubMed]
17. Antoniotti, V.; Spadaccini, D.; Ricotti, R.; Carrera, D.; Savastio, S.; Goncalves Correia, F.P.; Caputo, M.; Pozzi, E.; Bellone, S.; Rabbone, I.; et al. Adherence to the Mediterranean Diet Is Associated with Better Metabolic Features in Youths with Type 1 Diabetes. *Nutrients* **2022**, *14*, 596. [CrossRef] [PubMed]
18. Yiannakou, I.; Singer, M.R.; Jacques, P.F.; Xanthakis, V.; Ellison, R.C.; Moore, L.L. Adherence to a Mediterranean-Style Dietary Pattern and Cancer Risk in a Prospective Cohort Study. *Nutrients* **2021**, *13*, 4064. [CrossRef] [PubMed]
19. Augimeri, G.; Bonofiglio, D. The Mediterranean Diet as a Source of Natural Compounds: Does It Represent a Protective Choice against Cancer? *Pharmaceuticals* **2021**, *14*, 920. [CrossRef] [PubMed]
20. Mattavelli, E.; Olmastroni, E.; Bonofiglio, D.; Catapano, A.L.; Baragetti, A.; Magni, P. Adherence to the Mediterranean Diet: Impact of Geographical Location of the Observations. *Nutrients* **2022**, *14*, 2040. [CrossRef] [PubMed]

MDPI

Article

Impact of Mediterranean Diet Food Choices and Physical Activity on Serum Metabolic Profile in Healthy Adolescents: Findings from the DIMENU Project

Fabrizio Ceraudo [1], Giovanna Caparello [2,†], Angelo Galluccio [2,3,†], Ennio Avolio [2], Giuseppina Augimeri [1], Daniela De Rose [1], Adele Vivacqua [1], Catia Morelli [1], Ines Barone [1], Stefania Catalano [1,4], Cinzia Giordano [1,4], Diego Sisci [1,4] and Daniela Bonofiglio [1,4,*]

1 Department of Pharmacy, Health and Nutritional Sciences, University of Calabria, Rende, 87036 Cosenza, Italy; fabrizio.cer96@gmail.com (F.C.); giuseppina.augimeri@unical.it (G.A.); daniela.derose@unical.it (D.D.R.); adele.vivacqua@unical.it (A.V.); catia.morelli@unical.it (C.M.); ines.barone@unical.it (I.B.); stefcatalano@libero.it (S.C.); cinzia.giordano@unical.it (C.G.); diego.sisci@unical.it (D.S.)

2 Health Center srl, 87100 Cosenza, Italy; caparello.giovanna@gmail.com (G.C.); angelo.galluccio@yahoo.it (A.G.); ennioavolio@libero.it (E.A.)

3 Department of Clinical and Experimental Medicine, University Magna Graecia, 88100 Catanzaro, Italy

4 Centro Sanitario, University of Calabria, 87036 Rende, Italy

* Correspondence: daniela.bonofiglio@unical.it; Tel.: +39-0984-496208; Fax: +39-0984-496203

† These authors contributed equally to this work.

Citation: Ceraudo, F.; Caparello, G.; Galluccio, A.; Avolio, E.; Augimeri, G.; De Rose, D.; Vivacqua, A.; Morelli, C.; Barone, I.; Catalano, S.; et al. Impact of Mediterranean Diet Food Choices and Physical Activity on Serum Metabolic Profile in Healthy Adolescents: Findings from the DIMENU Project. *Nutrients* **2022**, *14*, 881. https://doi.org/10.3390/nu14040881

Academic Editor: Giuseppe Grosso

Received: 24 January 2022
Accepted: 18 February 2022
Published: 19 February 2022

Abstract: Adolescent nutrition and healthy dietary patterns, particularly the Mediterranean diet (MD), have been associated with improved health status and decreased risk of various chronic and metabolic diseases later in life. The aim of this study was to evaluate the impact of the Mediterranean food choices on lipid and glycemic metabolic profile in the total population and in adolescents grouped according to their physical activity (PA) levels at the time of recruitment (T0) and after six months from the administration of a personalized Mediterranean meal plan (T1). As part of the DIMENU study, 85 adolescents underwent measurements of lipid and glucose metabolic profile at T0 and T1. Using three positive items from KIDMED test related to the consumption of typical Mediterranean food (olive oil, fish, and nuts) and three negative items on dietary habits (going to fast-food, consuming biscuits, and candies), we categorized adolescents into six sets in which biochemical parameters were analyzed. In the total sample, significant reductions in serum total cholesterol, LDL, and glucose concentrations were observed for all the sets over the study period. Notably, active subjects, who had a better serum metabolic profile, showed significant improvements of glycemic control after 6 month follow up, while in sedentary adolescents and in those performing moderate PA significant reduction in glycemia, total cholesterol, and LDL was found in all sets. In conclusion, adopting the typical Mediterranean food choices led to a significant reduction in glucose and lipid profile in healthy adolescents, thus making the MD and PA a winning combination for health status.

Keywords: Mediterranean diet; adolescence; physical activity; metabolic profile

1. Introduction

The Mediterranean diet (MD) represents a nutritional model inspired by typical culinary traditions of populations bordering the Mediterranean basin, which share the same food availability. The identification of this geographical area is based on several scientific and epidemiological evidence demonstrated by Ancel Keys during the second half of the last century in the "Seven Country Study" [1]. Nowadays, the MD is considered a gold standard of healthy eating as far as it is related to greater longevity and improvement in health and quality of life [2], as well as a reduction of the risk of cardiovascular disease

(CVDs), cancer, obesity, diabetes, osteoporosis, and cognitive illnesses [3–9]. Benefits of the MD are due to the daily consumption of whole grains, in association with legumes, fruits, and vegetables as sources of vitamins and minerals, as well as low glycemic index carbohydrates and fiber, which can slow down the digestion of starch and the absorption of sugars with a lower increase in blood glucose over the postprandial period, improving insulin sensitivity and reducing the cholesterol absorption [10]. Soluble fiber, thanks to its ability to bind bile acids, also stimulates the growth of the gut microbiome (prebiotic effect), which ferments fibers with the production of short chain fatty acids, such as acetic, butyric, and propionic acids, and they are able to modulate the gluconeogenesis and lipogenesis in the liver and limit the production of potentially carcinogenic substances [11]. The daily consumption of extra virgin olive oil along with additional consumption of nuts are sources of antioxidant molecules, such as tocopherol (vitamin E), and unsaturated fatty acids (oleic and linoleic acids) that reduce the LDL cholesterol without affecting HDL [12,13]. It also involves moderate consumption of fish, the source of high biological value proteins and omega-3 fatty acids, which perform an anti-inflammatory function and a triglyceride-lowering activity, interfering with triglyceride incorporation in VLDL at hepatic level [14]. Regular but moderate consumption of wine, generally during meals, provides good amounts of polyphenols, including resveratrol, known for its antioxidant and anti-inflammatory effects, partly due to the ability to reduce cyclooxygenase-2 expression and consequently the prostaglandin production [15–17]. Weekly consumption of poultry and dairy products assures great protein sources with a moderate fat content. Occasional consumption of red meat and sweets is associated with a reduced intake of cholesterol and sugar [18,19]. The health effects of this pattern do not derive only from their individual components, but from the synergy between the MD components [20]. MD pattern is not just a way of eating, but it is a real dietary model that involves the total lifestyle and impacts on daily habits thanks to seasonality and biodiversity, accompanied by cultural elements such as conviviality, adequate night's rest, and physical activity [21].

Therefore, establishing and applying healthy eating behaviors starting from adolescence is important so that they persist into adulthood [22]. They represent the key elements, combined with regular physical activity, associated with a reduction in risk of obesity and in the prevention of metabolic syndrome and chronic diseases [23]. Adolescence is a period of physical growth and rapid development characterized by significant changes in cognitive, physiological, and emotional profile, with effects that can affect the quality of life, well-being, and health of individuals. In the recent decades, the phenomenon of food "westernization" has been described as the consumption of foods rich in refined carbohydrates, saturated fats, salt, and proteins (chips, salty snacks, fast food, and candies) and soft drinks. In addition, the sedentary lifestyle induced by modern society caused by automated equipment, motorized transport, and the increased time spent watching television is responsible for the increased incidence of obesity as well as metabolic and chronic non-communicable diseases [24].

The aim of this study was to evaluate the impact of the MD food choices on lipid and glycemic metabolic profile in the total population and in adolescents grouped according to their physical activity (PA) levels at the time of recruitment (T0) and after six months from the administration of a personalized Mediterranean meal plan (T1).

2. Materials and Methods

2.1. Study Population

The adolescents were recruited into the DIMENU research project (DIeta MEditerranea e NUoto, FESR-FSE 2014–2020. Prot 52243/2017) enrolled by the Castrolibero Institute of Education (Cosenza, Italy) and by sports associations of the Calabria region, Italy [25–28]. The total population studied is composed of 85 subjects (44 girls and 41 boys) aged between 14 and 17 years. Criteria of exclusion from the study included health problems, drug use, and any type of restrictive diet (i.e., low calorie, low carbohydrate, and low fat content). The participants did not have any kind of cognitive or physical/motor limitation. The adolescents participating in the study provided verbal informed consent, while their parents signed a written informed consent form to allow their participation. The rationale of the research project and the adequacy of the protocol were approved by the Ethics Committee of the University of Calabria, Italy (# 5727/2018).

2.2. Nutritional History

The assessment of the participants' nutritional status and medical history was made during an interview with a team of professionals (endocrinologists and biologist nutritionists). During the interview, the students provided the anamnestic data (general data, medical history, eating habits, and intensity of physical activity). After that, participants were divided into three groups, on the basis of their difference in the physical activity levels, expressed in MET (metabolic-equivalent unit expressed as 1 kcal/kg/hour), in accordance with recommendations issued by the WHO [29]. Thus, the groups obtained were: (1) Group A, physical inactivity, less than 3 METs; (2) Group B, moderate physical activity, between 3 and 6 METs (cycling, dancing, brisk walking, gymnastics, ballet, water aerobics, recreational swimming) for at least 60 min a day; (3) Group C, vigorous physical activity, above 6 METs (jogging or running, boxing, tennis, football, basketball, squash, swimming, aerobic dance and volleyball) for at least 60 min per day.

2.3. Mediterranean Diet Adherence Test (KIDMED)

The KIDMED test (Mediterranean Diet Quality Index for children and teenagers) has been used to assess adherence to Mediterranean dietary patterns in children and adolescents [30], updated on the new food pyramid of the International Foundation of MD, as previously reported [26]. The questionnaire is made up of 16 questions, of which 4 denote a negative connotation and 12 questions denote a positive connotation to MD. To evaluate the impact of the Mediterranean food choices on the different parameters analyzed at T0 and T1, we selected questions with positive connotations from KIDMED test, which were categorized in the following sets: set 1: "Do you consume olive oil every day?"; set 2: "Do you consume olive oil every day?" and "Do you consume $\geq$ 2 portions of fish/per week?"; set 3: "Do you consume olive oil every day?", "Do you consume $\geq$ 2 portions of fish/per week?", and "Do you consume $\geq$ 2 servings of nuts per week?". Moreover, on the basis of negative items impacting the adherence to the MD from KIDMED test, we categorized three sets: set 4:"Do you go to fast-food restaurants more than once a week?"; set 5: "Do you go to fast-food restaurants more than once a week?" and "Do you consume biscuits or baked goods for breakfast?"; set 6: "Do you go to fast-food restaurants more than once a week?", "Do you consume biscuits or baked goods for breakfast?", and "Do you consume sweets and candies every day?" (Figure 1).

2.4. Anthropometric Parameters and Bioelectrical Impedance Analysis

Anthropometric data were collected using a validated protocol [31]. Participants' weights were determined using the KERN MPC250K100 M scale with a load capacity of 250 kg and an accuracy of 100 g. Height was determined using a Seca stadiometer with a maximum capacity of 220 cm and an accuracy of 1 mm. The body circumferences of each participant were measured using a validated ergonomic tape Seca 201, with a measuring range from 1 to 205 cm and a division of 1 mm. Body mass index (BMI) was calculated by dividing the body weight in kilograms by the square of height in meters [BMI = weight (kg)/height^2 (m)]. The BMI z-score was calculated on the basis of the World Health Organization data [BMI z-score = $[(\mathrm{BMI}/\mathrm{M}(t))\mathrm{L}(t) - 1]/\mathrm{L}(t)\mathrm{S}(t)$]. In particular, waist and hip circumferences were used to calculate the waist/hip ratio (WHR). Body composition assessment was performed after a 12 h overnight fast according to the measurement protocol using bioimpedentiometric analysis (BIA) (single-frequency 50 kHz BIA 101 S, Akern-Systems, Florence, Italy). Each subject underwent BIA, performed to evaluate resistance, reactance, phase angle (PhA), total body water (TBW), body cell mass (BCM), fat-free mass (FFM), and fat mass (FM). Data were analyzed using Bodygram Plus software

Version 1.2.2.8. (Akern Srl; Florence, Italy). Measurements and assessments of bioelectrical parameters were made at recruitment time (T0) and after six months (T1).

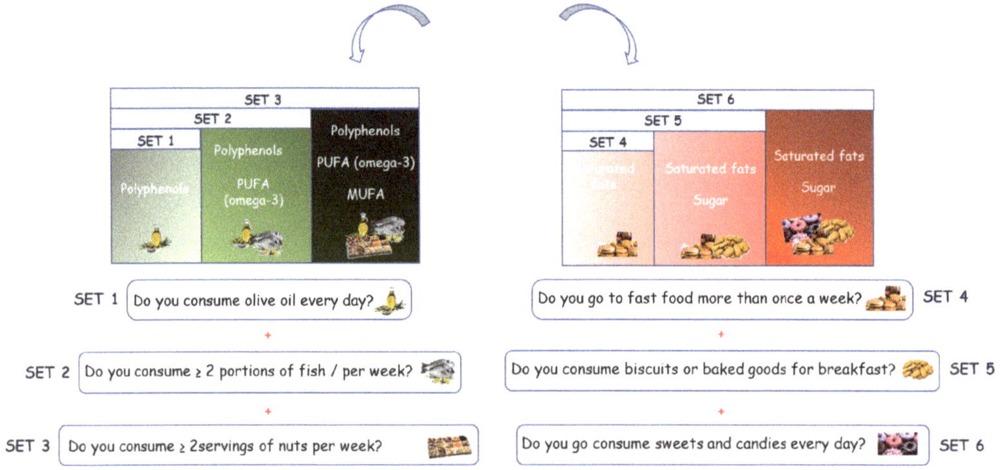

Figure 1. Classification of the Mediterranean Diet food choices according to selected questions with positive (sets 1, 2 and 3) or negative (sets 4, 5, and 6) impact on MD adherence.

2.5. Biochemical Measurements

Venous blood samples were collected after 8–10 h of overnight fasting, both at baseline (T0) and T1. The serum was obtained after centrifugation at 3000 rpm for 10 min and stored in sterile tubes at 4 °C for no more than 4 h during the morning of collection. The biochemical parameters were determined by a Konelab 20i chemistry analyzer (Thermo Electron Corporation, Vantaa, Finland) according to standard procedures. Subsequently, the serum samples were stored at −80 °C. Serum insulin levels were measured with an enzyme-linked immunosorbent assay (Novatec Immundiagnostica GmbH, Dietzenbach, Germany) following the manufacturer's instructions. The lowest detectable insulin concentration was 0.25 µIU/mL at a 95% confidence limit; the intra-assay variability was within ≤5%.

2.6. Mediterranean Personalized Food Plan

Each participant received a personalized Mediterranean plan that was based on their nutritional status and different levels of physical activity (PA). Throughout the program period, nutritionists provided participants with indications on the choice of typical Mediterranean foods. The dietary approach is based on the latest guidelines of the MD as previously reported [25]. The dietary scheme provided 15–20% of calories through protein, 45–60% of calories through carbohydrates, and 25–30% of calories through fat, with the redistribution of macro- and micronutrients according to the different daily energy expenditure (TDEE) of each subject as recommended by the Italian Society of Human Nutrition [32]. Moreover, we have to underline the fact that distribution of some macronutrients such us protein (range from 1 to 1.5 g/day) were also customized according to physical activity (PA) level (sedentary, moderately active, and vigorous), and thus the same diet plan was not provided to everyone. The foods included in the diet are obviously typically Mediterranean ones, and meals included an abundance of plant food (fruits, vegetables, whole grains, nuts, and legumes); low-fat dairy products, fish, poultry, and eggs in moderate amounts; olive oil as the primary source of fat; and low consumption of red meats, processed foods, and saturated lipids. Meals and food plans were designed using MetaDieta software version 4.2.1. (Meteda S.r.l, Roma, Italy).

2.7. Statistical Analysis

All statistical analyses were performed using SigmaPlot Version 12.0 (Systat, San Jose, CA, USA). A Kolmogorov–Smirnov test (with Lilliefors' correction) was used to verify data normality. Data were reported as the mean and standard deviation (SD), and statistical differences between samples were evaluated by using parametric tests (one-way ANOVA and Student's t-test). Statistical significance was set at $p < 0.05$.

3. Results

3.1. General Characteristics and Metabolic Profile of the Study Population

Anthropometric and bioimpedance measurements as well as metabolic parameters were evaluated in the total sample of adolescents (n = 85) before they started the food plan (T0) and 6 months after (T1) as reported in Table 1. From the comparison of the total sample in the two observation periods, there were no significant changes, except for the BMI, which was increased at T1 (p = 0.0237), remaining within the range of normal values. Interestingly, after 6 months, a significant reduction in fasting blood glucose (p = 0.0001), total cholesterol (p = 0.0002), and LDL (p = 0.0009) was found.

Table 1. Anthropometric and bioimpedance parameters and metabolic profile in adolescents at T0 and T1.

	T0		T1		
Anthropometric Parameters	**Mean**	**SD**	**Mean**	**SD**	***p*-Value**
Weight (kg)	62.426	12.39	63.518	12.483	0.5685
Height (cm)	165.854	7.818	167.134	8.508	0.3087
BMI (kg/m^2)	22.685	3.685	23.75	2.252	**0.0237**
BMI z-score	0.48	0.86	0.53	0.76	0.6884
WHR	0.772	0.047	0.781	0.065	0.3024
Bioimpedentiometric Parameters					
PhA (°)	6.128	0.693	6.285	0.777	0.1860
BCM (Kg)	26.468	5.714	26.929	5.844	0.6043
FFM (Kg)	48.627	8.987	48.866	9.144	0.8632
FM (Kg)	13.799	7.448	15.291	7.057	0.1826
TBW (%)	36.241	6.587	36.015	6.962	0.8327
Metabolic Profile					
Total cholesterol (mg/dL)	155.24	27.7	139.89	24.31	**0.0002**
LDL (mg/dL)	83.4	25.96	71.576	19.02	**0.0009**
HDL (mg/dL)	58.93	14.19	56.76	13.42	0.3071
Triglycerides (mg/dL)	64.39	31.07	57.67	24.43	0.1189
Glucose (mg/dL)	83.46	7.47	77.41	8.48	**0.0001**
Insulin (µIU/mL)	10.35	4.89	11.03	6.1	0.4237

BMI, body mass index; WHR, waist hip ratio; PhA, phase angle; BCM, body cell mass; FFM, fat-free mass; FM, fat mass; TBW, total body water; LDL, low-density lipoprotein; HDL, high-density lipoprotein. Statistical differences were determined by Student's t-test. In bold are reported statistically significant values.

3.2. Impact of the Mediterranean Diet Food Choices on the Adolescent Metabolic Profile

Having previously evaluated that in this population the KIDMED score at T1 compared with T0 were significantly increased after 6 month follow-up (T0 = 6.04 $\pm$ 2.34 vs. T1 = 6.94 $\pm$ 2, p = 0.006) [25], in order to deeper investigate the contribution of the MD food choices in the improvement of metabolic profile, we selected questions from the KIDMED test relating to the consumption of foods affecting lipid profile (total cholesterol, LDL, HDL, and triglycerides) and glucose homeostasis (glucose and insulin). Table 2 shows the changes in the metabolic profile of the participants categorized into three sets, which refer to some of the questions chosen from the KIDMED test as follows: (1) "Do you consume olive oil every day?"; (2) "Do you consume olive oil every day?" and "Do you consume $\geq$2 portions of fish/per week?"; (3) "Do you consume olive oil every day?", "Do you consume

≥2 portions of fish/per week?", and "Do you consume ≥2 servings of nuts per week?". Analyzing the biochemical parameters in the sets identified, at both times of observation, we found that particular significance emerged. Interestingly, we observed significant improvements in fasting blood glucose in all the sets (p = 0.0001 in set 1, p = 0.0001 in set 2, and p = 0.0019 in set 3). Total cholesterol and LDL concentrations significantly decreased in set 1 (p = 0.0004 and p = 0.0012, respectively), in set 2 (p = 0.0173 and p = 0.0123, respectively), and in set 3 (p = 0.0022 and p = 0.0086, respectively) (Table 2).

Table 2. Comparison between T0 and T1 in the metabolic parameters of adolescents categorized in three sets according to the positive KIDMED items.

KIDMED Items		**SET 1**			**SET 2**			**SET 3**		
Subjects		**(78 vs. 78)**			**(51 vs. 53)**			**(20 vs. 17)**		
Parameters		**Mean**	**SD**	***p*-Value**	**Mean**	**SD**	***p*-Value**	**Mean**	**SD**	***p*-Value**
Total cholesterol (mg/dl)	T0 T1	155.00 140.08	27.30 23.92	**0.0004**	153.82 141.77	28.36 22.15	**0.0173**	159.45 133.35	26.72 20.06	**0.0022**
LDL (mg/dl)	T0 T1	83.28 71.44	25.56 18.92	**0.0012**	82.84 72.00	26.26 16.12	**0.0123**	89.30 71.47	21.22 17.02	**0.0086**
HDL (mg/dl)	T0 T1	59.04 57.08	14.25 13.44	0.3782	57.95 57.77	15.06 14.61	0.9508	57.05 50.94	13.44 11.57	0.1511
Triglycerides (mg/dl)	T0 T1	63.31 57.82	27.21 24.11	0.1843	65.07 59.66	35.33 24.67	0.3659	65.15 54.47	26.66 17.86	0.1691
Glucose (mg/dl)	T0 T1	83.46 77.18	7.40 8.51	**0.0001**	83.84 75.79	7.79 7.10	**0.0001**	83.84 75.59	7.79 6.97	**0.0019**
Insulin (µIU/mL)	T0 T1	10.42 10.63	5.08 5.55	0.8056	9.42 10.26	4.69 6.00	0.3315	9.42 12.02	4.69 7.69	0.2151

LDL, low-density lipoprotein; HDL, high-density lipoprotein. Statistical differences were determined by Student's *t*-test. In bold are reported statistically significant values.

Moreover, questions from KIDMED test relating to improper eating habits were used to categorize adolescents that gave a negative answer to the questions in sets 4, 5, and 6, as described: (4)"Do you go to fast-food restaurants more than once a week?"; (5) "Do you go to fast-food more than once a week?" and "Do you consume biscuits or baked goods for breakfast?"; (6) "Do you go to fast-food more than once a week?", "Do you consume biscuits or baked goods for breakfast?", and "Do you consume sweets and candies every day?". Notably, total cholesterol and LDL levels significantly decreased in set 4 (p = 0.0011 and p = 0.0047, respectively), in set 5 (p = 0.0026 and p = 0.0226, respectively), and in set 6 (p = 0.0049 and p = 0.0334, respectively). Furthermore, there was a drastic reduction of fasting glycaemia in all the sets (p = 0.0001 in set 4, p = 0.0001 in set 5, and p = 0.0024 in set 6) (Table 3).

3.3. Impact of the Mediterranean Diet Food Choices on Metabolic Profile in the Adolescents Grouped According to the Different Physical Activity Levels

On the basis of the self-reported PA intensity levels, we grouped our adolescents into the physical inactivity (Group A; n = 23), moderate-intensity PA (Group B; n = 34), and vigorous-intensity PA (Group C; n = 28) levels, which were confirmed by interview over the study period [25]. We have previously reported that the MD adherence increased particularly in adolescents performing moderate (KIDMED score at T0: 5.57 ± 2.30 and at T1: 6.94 ± 1.66) and vigorous (KIDMED score at T0: 6.30 ± 2.16 and at T1: 7.61 ± 1.77) PA levels compared to sedentary (KIDMED score at T0: 5.96 ± 2.10 and at T1: 6.61 ± 2.35) [14]. Here, we analyzed, in the three PA groups of adolescents, the intra- and inter-group differences in the metabolic profile at T0 and T1 (Table 4). From the intragroup analysis, we found a significant reduction in fasting blood glucose (p = 0.0052), total cholesterol (p = 0.0008), and LDL (p = 0.0028) in Group A. Adolescents from Group B showed total cholesterol and LDL levels that were significantly reduced (p = 0.0001 and p = 0.0152, respectively). In Group C, subjects had significant reductions in fasting glucose (p = 0.0001). Table 4 also shows the intergroup comparison (A vs. B, A vs. C, and B vs. C) at T0 and T1, performed by using ANOVA test. At T0, Group C had statistically higher fasting blood glucose values (p = 0.0309) compared to Group B, while plasma LDL concentrations were statistically lower in Groups B and C than in Group A (p = 0.0488 and p = 0.0015, respectively). Insulinemia was lower in Group B than in Group A (p = 0.0277) at T0, while both Groups B and C had significantly lower values than Group A (p = 0.0009 and p = 0.0001, respectively) at T1.

Table 3. Comparison between T0 and T1 in the metabolic parameters of adolescents categorized in three sets according to the negative KIDMED items.

KIDMED Items		**SET 4**			**SET 5**			**SET 6**		
Subjects		**(73 vs. 77)**			**(55 vs. 68)**			**(30 vs. 31)**		
Parameters		**Mean**	**SD**	***p*-Value**	**Mean**	**SD**	***p*-Value**	**Mean**	**SD**	***p*-Value**
Total cholesterol (mg/dl)	T0 T1	153.96 139.87	27.28 24.53	**0.0011**	154.44 140.19	26.08 25.13	**0.0026**	155.57 136.29	25.92 25.58	**0.0049**
LDL (mg/dl)	T0 T1	82.47 71.79	26.12 18.99	**0.0047**	81.40 72.07	25.42 19.35	**0.0226**	79.40 67.58	23.28 19.22	**0.0344**
HDL (mg/dl)	T0 T1	58.41 56.61	14.51 13.65	0.4348	60.07 56.51	14.95 14.16	0.1791	63.07 56.13	16.08 16.45	0.1012
Triglycerides (mg/dl)	T0 T1	65.45 57.23	33.15 24.79	0.0865	64.80 57.91	34.96 25.26	0.2076	65.63 63.26	27.59 28.03	0.7400
Glucose (mg/dl)	T0 T1	83.92 77.87	7.00 8.47	**0.0001**	84.00 77.85	6.44 8.76	**0.0001**	82.73 77.13	6.28 7.45	**0.0024**
Insulin (μIU/mL)	T0 T1	10.66 10.76	4.78 5.99	0.9079	10.58 10.80	4.95 6.26	0.8276	9.76 11.60	3.68 6.07	0.1587

LDL, low-density lipoprotein; HDL, high-density lipoprotein. Statistical differences were determined by Student's *t*-test. In bold are reported statistically significant values.

In addition, the impact of specific MD food choices over time (T0 vs. T1) was also observed for the three groups on the basis of the PA performed (Table 5). The results in Group A demonstrate a significant decrease in total cholesterol and LDL in set 1 (p = 0.0032 and p = 0.0057, respectively) and in fasting blood glucose (p = 0.0019). In set 2, significant reductions were recorded for total cholesterol (p = 0.0328) and blood glucose (p = 0.0070). In Group B, a significant reduction in total cholesterol and LDL was present in set 1 (p = 0.0135 and p = 0.0059, respectively) and in glucose levels in set 2 (p = 0.0392). Regarding Group C, in all sets, fasting glucose was significantly reduced (p = 0.0001 in set 1, p = 0.0001 in set 2, p = 0.0019 in set 3). No significant changes were found in these sets for both anthropometric and bioimpedance parameters (data not shown).

Table 4. Biochemical and metabolic parameters in adolescents grouped according to different PA levels (Groups A, B, and C) at T0 and T1.

	Group A				Group B				Group C									
	T0		T1		T0		T1		T0		T1		*p*-Value T0			*p*-Value T1		
	Mean	**SD**	**Mean**	**SD**	**Mean**	**SD**	**Mean**	**SD**	**Mean**	**SD**	**Mean**	**SD**	**A vs. B**	**A vs. C**	**B vs. C**	**A vs. B**	**A vs. C**	**B vs. C**
Total cholesterol (mg/dl)	164.61	28.51	138.65 *	19.65	154.85	29.25	**139.29** *	26.32	148.00	23.46	141.64	25.13	0.3854	0.0835	0.5885	0.9948	0.9021	0.9256
LDL (mg/dl)	97.87	30.08	74.35 *	17.09	82.09	23.26	**68.62** *	21.26	73.11	20.19	72.89	16.98	**0.0488**	**0.0015**	0.3238	0.5094	0.9604	0.6557
HDL (mg/dl)	54.65	12.58	52.48	11.98	58.53	15.84	58.88	15.06	62.93	12.58	57.71	11.43	0.5621	0.0955	0.4375	0.1822	0.3470	0.9370
Triglycerides (mg/dl)	60.70	21.10	59.13	21.68	70.50	39.47	59.21	28.13	60.00	25.64	54.61	21.10	0.4741	0.9965	0.3849	>0.9999	0.7914	0.7454
Glucose (mg/dl)	82.61	6.34	76.57 *	7.55	81.62	8.35	77.94	10.18	86.39	6.45	77.46 *	6.60	0.8685	0.1585	**0.0309**	0.823	0.9262	0.9741
Insulin (µIU/mL)	12.43	6.04	15.58	8.00	9.04	4.24	10.04	3.92	10.20	4.07	8.51	4.19	**0.0277**	0.2243	0.6109	**0.0009**	**<0.0001**	0.5181

PA, physical activity; LDL, low-density lipoprotein; HDL, high-density lipoprotein. Statistical differences between T0 and T1 in each PA group were determined by Student's *t*-test. * $p < 0.005$. Statistically significant differences among Groups A, B, and C at T0 and T1 were calculated by ANOVA test. In bold are reported statistically significant values.

Table 5. Biochemical and metabolic parameters in adolescents from Groups A, B and C categorized into three sets according to the positive KIDMED items at T0 and T1.

KIDMED		SET 1			SET 2			SET 3		
Subject Group A		**(22 vs. 21)**			**(12 vs. 11)**			**(4 vs. 5)**		
Parameters		**Mean**	**SD**	***p*-Value**	**Mean**	**SD**	***p*-Value**	**Mean**	**SD**	***p*-Value**
Total cholesterol (mg/dl)	T0 T1	163.82 139.90	28.92 20.24	**0.0032**	163.42 137.64	30.66 22.35	**0.0328**	144.33 128.20	20.77 22.33	0.3039
LDL (mg/dl)	T0 T1	75.19 97.36	17.73 30.69	**0.0057**	74.73 96.25	15.72 32.92	0.0737	82.00 75.80	24.60 16.33	0.6626
HDL (mg/dl)	T0 T1	54.18 53.24	12.67 12.40	0.8071	55.25 50.82	13.50 14.68	0.4592	48.83 39.40	15.59 7.57	0.2687
Triglycerides (mg/dl)	T0 T1	61.50 57.52	21.23 18.90	0.5205	60.08 60.00	22.17 13.13	0.9918	65.83 64.80	15.97 12.95	0.9176
Glucose (mg/dl)	T0 T1	82.95 75.90	6.26 7.63	**0.0019**	83.17 73.91	6.78 8.08	**0.0070**	83.17 75.80	6.78 9.98	0.2497
Insulin (µIU/mL)	T0 T1	15.58 14.41	8.18 7.15	0.6209	15.39 14.98	8.73 9.02	0.9080	19.45 11.03	4.78 9.12	0.2151
Subject Group B		**(30 vs. 31)**			**(20 vs. 22)**			**(5 vs. 4)**		
Parameters		**Mean**	**SD**	***p*-Value**	**Mean**	**SD**	***p*-Value**	**Mean**	**SD**	***p*-Value**
Total cholesterol (mg/dl)	T0 T1	155.53 137.97	27.93 25.89	**0.0135**	151.75 141.95	31.21 18.23	0.2266	149.00 134.250	38.33 18.428	0.5062
LDL (mg/dl)	T0 T1	83.10 67.16	22.35 21.24	**0.0059**	82.40 70.09	24.55 16.90	0.0636	83.80 72.750	37.83 25.786	0.6351
HDL (mg/dl)	T0 T1	58.63 58.87	15.83 15.18	0.9520	54.90 58.91	16.98 15.91	0.4341	53.80 50.250	11.71 7.182	0.6138
Triglycerides (mg/dl)	T0 T1	60.03 68.27	29.27 32.17	0.3411	64.82 71.20	29.91 35.19	0.4705	56.80 57.000	33.84 17.010	0.9918
Glucose (mg/dl)	T0 T1	81.10 77.94	8.01 10.34	0.1882	81.50 76.27	8.31 7.59	**0.0392**	81.50 72.75	8.31 8.34	0.1367
Insulin (µIU/mL)	T0 T1	9.04 10.06	4.51 3.92	0.3492	7.80 9.86	4.56 4.14	0.1327	7.80 12.55	4.56 3.70	0.1610
Subject Group C		**(26 vs. 26)**			**(19 vs. 20)**			**(11 vs. 8)**		
Parameters		**Mean**	**SD**	***p*-Value**	**Mean**	**SD**	***p*-Value**	**Mean**	**SD**	***p*-Value**
Total cholesterol (mg/dl)	T0 T1	146.92 142.73	23.48 24.84	0.5348	149.58 143.85	22.04 26.41	0.4678	151.73 136.125	19.98 21.464	0.1215
LDL (mg/dl)	T0 T1	71.58 73.50	18.00 16.45	0.6898	73.68 72.60	17.42 16.01	0.8412	73.64 68.125	19.13 14.066	0.5001
HDL (mg/dl)	T0 T1	63.62 58,04	12.55 11,80	0.1049	64.32 60,35	12.86 12,44	0.2770	65.36 58,56	14.04 8.73	0.2492
Triglycerides (mg/dl)	T0 T1	59.12 55.42	25.50 21.56	0.5850	58.32 53.80	27.82 22.80	0.5363	64.18 46.750	32.79 19.002	0.1972
Glucose (mg/dl)	T0 T1	86.62 77.46	6.64 6.60	**0.0001**	86.74 76.30	7.23 6.11	**0.0001**	86.74 76.88	7.23 4.22	**0.0019**
Insulin (µIU/mL)	T0 T1	10.19 8.51	4.21 4.19	0.1555	10.03 8.11	4.49 4.32	0.1817	10.03 7.10	4.49 3.81	0.2151

LDL, low-density lipoprotein; HDL, high-density lipoprotein. Group C, vigorous physical activity. Statistical differences were determined by Student's *t*-test. In bold are reported statistically significant values.

Differences between T0 and T1 were also performed in the three groups on the basis of PA (A, B, and C) (Table 6). Again, significant differences emerged regarding the lipid metabolic profile and glucose concentrations. Adolescents from Group A had a reduction in total cholesterol and LDL in all three sets ($p = 0.0014$ and $p = 0.0052$, in set 4; $p = 0.0066$ and $p = 0.0257$ in set 5; $p = 0.0024$ and $p = 0.0251$ in set 6), while fasting glucose was significantly reduced in sets 4 and 6 ($p = 0.0052$ and $p = 0.0034$, respectively). Moreover, adolescents from Group B showed significant changes in total cholesterol and LDL in set 4 ($p = 0.0454$ and $p = 0.0437$, respectively) and in set 6 ($p = 0.0264$ and 0.0473, respectively), and changes in total cholesterol in set 5 ($p = 0.0406$). Interestingly, in Group C, subjects had significant reductions in fasting blood glucose in the three sets ($p = 0.0001$ in set 4, $p = 0.0001$ in set 5, and $p = 0.0054$ in set 6), along with a reduction in insulinemia in sets 4 and 5 ($p = 0.0274$ and $p = 0.0338$, respectively). No significant differences emerged with regards to the anthropometric and bioimpedance parameters (data not shown).

Table 6. Biochemical and metabolic parameters in adolescents from Groups A, B and C categorized into three sets according to the negative KIDMED items at T0 and T1.

KIDMED		SET 4			SET 5			SET 6		
Subject Group A		**(21 vs. 22)**			**(17 vs. 21)**			**(7 vs. 10)**		
Parameters		**Mean**	**SD**	***p*-Value**	**Mean**	**SD**	***p*-Value**	**Mean**	**SD**	***p*-Value**
Total cholesterol (mg/dl)	T0 T1	160.38 137.09	25.18 19.09	**0.0014**	156.76 136.57	23.80 19.40	**0.0066**	161.14 128.60	20.71 16.22	**0.0024**
LDL (mg/dl)	T0 T1	94.62 73.32	28.96 17.16	**0.0052**	89.94 72.71	27.97 17.34	**0.0257**	91.71 66.00	29.47 12.42	**0.0251**
HDL (mg/dl)	T0 T1	53.67 52.14	12.68 12.42	0.6915	55.47 52.24	12.96 12.72	0.4449	57.57 50.40	16.28 13.19	0.3317
Triglycerides (mg/dl)	T0 T1	60.86 58.18	22.10 22.20	0.6942	57.41 58.10	20.05 22.75	0.9232	61.14 60.70	24.33 23.31	0.9703
Glucose (mg/dl)	T0 T1	83.29 77.09	6.20 7.46	**0.0052**	83.71 76.71	6.02 7.43	**0.0034**	82.43 75.30	4.58 8.23	0.0566
Insulin (μIU/mL)	T0 T1	12.66 15.14	6.28 8.09	0.2699	13.00 15.06	6.69 8.28	0.4122	10.94 16.40	3.43 7.77	0.1042
Subject Group B		**(29 vs. 29)**			**(22 vs. 27)**			**(11 vs. 12)**		
Parameters		**Mean**	**SD**	***p*-Value**	**Mean**	**SD**	***p*-Value**	**Mean**	**SD**	***p*-Value**
Total cholesterol (mg/dl)	T0 T1	155.41 139.39	30.90 28.67	**0.0454**	158.32 141.15	28.29 28.44	**0.0406**	168.64 140.73	26.09 29.63	**0.0264**
LDL (mg/dl)	T0 T1	81.79 69.04	24.43 22.63	**0.0437**	82.82 70.88	22.78 22.32	0.0714	87.09 67.82	19.80 23.67	**0.0473**
HDL (mg/dl)	T0 T1	58.93 58.61	16.51 16.04	0.9399	60.59 58.04	16.40 16.18	0.5877	66.27 60.64	17.98 20.53	0.4931
Triglycerides (mg/dl)	T0 T1	72.97 59.04	42.15 30.35	0.1542	74.09 61.19	44.48 30.44	0.2355	75.18 62.64	30.10 39.01	0.4009
Glucose (mg/dl)	T0 T1	81.97 79.04	8.29 10.67	0.2480	82.50 79.46	7.28 10.97	0.2710	78.73 77.45	6.12 8.31	0.6824
Insulin (μIU/mL)	T0 T1	9.11 9.50	4.23 3.74	0.7112	8.52 9.56	3.41 3.87	0.3300	8.05 9.58	3.52 3.39	0.3027

Table 6. *Cont.*

KIDMED		SET 4			SET 5			SET 6		
Subject Group C		(23 vs. 26)			(16 vs. 20)			(12 vs. 9)		
Parameters		**Mean**	**SD**	***p*-Value**	**Mean**	**SD**	***p*-Value**	**Mean**	**SD**	***p*-Value**
Total cholesterol (mg/dl)	T0 T1	146.26 143.23	23.22 24.70	0.6615	146.63 143.40	25.11 27.03	0.7159	140.33 140.44	21.80 30.50	0.9923
LDL (mg/dl)	T0 T1	72.22 73.62	21.45 16.79	0.7994	70.38 73.15	23.43 18.53	0.6938	65.17 69.00	14.90 22.49	0.6434
HDL (mg/dl)	T0 T1	62.09 58.62	12.64 11.37	0.3165	64.25 59.50	14.34 12.45	0.2951	63.33 58.00	14.60 14.46	0.4158
Triglycerides (mg/dl)	T0 T1	60.17 54.54	27.71 21.15	0.4246	59.88 53.55	31.40 21.24	0.4767	59.50 67.67	26.75 19.89	0.4517
Glucose (mg/dl)	T0 T1	86.96 77.35	4.78 6.79	**0.0001**	86.38 77.05	5.21 6.96	**0.0001**	86.58 78.89	5.12 6.11	**0.0054**
Insulin (µIU/mL)	T0 T1	10.78 8.51	3.04 3.85	**0.0274**	10.83 8.06	3.37 3.98	**0.0338**	10.63 9.10	3.65 3.60	0.3497

LDL, low-density lipoprotein; HDL, high-density lipoprotein. Statistical differences were determined by Student's *t*-test. In bold are reported statistically significant values.

4. Discussion

In this study, we evaluated the impact of different MD food choices on serum metabolic parameters in a population of healthy adolescents performing different PA levels. Nowadays, the importance of optimal adherence to the MD and PA is widely known as the main tool in countering the onset of chronic non-communicable diseases, often associated with increased consumption of unhealthy foods (junk food) and an increasingly sedentary lifestyle. For this reason, it is necessary to promote the Mediterranean pattern, especially during adolescence, in order to educate the new generations to have good eating habits that will result in the maintenance of good health and a long life expectancy. Globally, adolescents show a poor adherence to nutritional recommendations, preferring an excess of energy coming from fats at the expense of that taken from carbohydrates and proteins; in addition, there is a low consumption of those foods that characterize MD such as fruits, vegetables, legumes, and fish, along with a widespread habit of skipping breakfast [25,26,28]. The daily consumption of sweets and sugary drinks affects a non-negligible share of adolescents [33]. In our population, the adherence to the MD was evaluated by the KIDMED test, which showed an increased score over the study period. We found a significant difference only for the BMI values, but not for the BMI z-score, in the follow-up study, although, in presence of unchanged other anthropometric and bioimpedance measurements, BMI was not the diagnostic measure for adiposity in adolescents. Conversely, it has been reported that BMI and waist circumference could be considered diagnostic tests for fatness, while WHR is less useful in adolescents [34]. Interestingly, we observed a significant reduction in the serum levels of glucose as well as total cholesterol and LDL in the total sample of adolescents, indicating that the promotion of MD pattern and the increased MD adherence had beneficial effects on serum metabolic profile. Specifically, when we categorized adolescents on the basis of specific MD food choices, we observed over the study period significant decreased levels of total cholesterol, LDL, and glycemia in the sets of participants who consumed "extra virgin olive oil every day" alone (set 1) or together with "two or more portions of fish per week" (set 2) and in combination with "two or more servings of nuts per week" (set 3). These questions referred to the intake of typical recommended MD foods that are rich with monounsaturated fats (MUFA), polyphenols, and polyunsaturated fatty acids (PUFA) including omega-3-fatty acids, representing essential Mediterranean diet components. Similarly, in the sets of participants who "don't go to fast-food restaurants more than once a week" alone (set 4) or in combination with "don't consume biscuits or baked goods for breakfast" (set 5) or with "don't consume sweets and candies every day" (set 6), we found in the follow-up reduced concentrations of serum total cholesterol, LDL, and glycemia. These results fit well with a recent open label study [35], in which Velázquez-López et al. reported that a Mediterranean pattern-based diet improves lipid and glycemic profile in obese children and adolescents.

Our previous observations showing that PA intensity levels positively influenced healthy dietary pattern [25] represent encouraging results since those who practice vigorous PA began to adopt good eating habits that could last over time. This is in line with other studies [28,36,37] in which the increase in consumption of junk food, sweets, and candies are related to the habits of mainly sedentary adolescents. It was interesting to note that in our population sample, active adolescents display a better metabolic profile compared to sedentary adolescents in terms of LDL and insulin concentrations. This leads us to speculate that the myokines, such as interleukin-6 (IL-6), produced in working skeletal muscle in larger amounts, play a beneficial role in metabolic homeostasis, improving insulin sensitivity and lipolysis and thus making the MD and PA a winning combination for health status. Notably, the blood chemistry parameters significantly improved with the adherence to the specific MD food consumption over the study period in all groups of adolescents from sedentary to moderate/vigorous PA intensity levels. In particular, in sedentary adolescents categorized by positive and negative items of the KIDMED test, we observed a significant reduction in the serum levels of total cholesterol, LDL, and glycemia over the study period, indicating the beneficial role of dietary pattern. A similar trend was found in adolescents who practice moderate PA, while in active subjects, glycemia was significantly reduced in the positive sets, and both glycaemia and insulinemia were decreased in the negative sets after 6-month follow-up. These latter findings fit well with the results from several meta-analyses [38–40] that demonstrated how MD guarantees optimal glycemic control when compared with other dietary models.

The limitations of the study include a relatively small sample size, particularly when the total sample was divided into the PA groups, as well as the lack of objective methods to measure PA levels. Adolescence is a phase of life that involves a series of complex alterations at endocrine levels, and this can represent confounding effects on hormonal and metabolic changes occurring during puberty. However, our study strengthens the importance of improving healthy dietary habits and encouraging our adolescents towards better food choices that have been pursued and confirmed in the follow-up. Future studies with a larger number of participants and a different age range (from adolescence to adulthood) and, in particular, the comparison among participants of different age could add considerable insights to the field.

5. Conclusions

In conclusion, findings from this study show that the introduction of a personalized food plan based on MD principles in an adolescent sample led to a significant improvement in glucose and lipid profile in the follow-up, particularly when subjects adopt to the typical MD food choices. This demonstrates the effectiveness of food education programs and how these are implemented and translated into good eating habits in the adolescent population and for the entire lifetime.

Author Contributions: F.C.—data collection, study design, and writing the manuscript; E.A., A.G., G.C. and F.C.—nutritional analyses; D.D.R. and A.V.—blood analyses; G.A., C.M., I.B., S.C. and C.G.—analyses and interpretation of data; D.S.—conception of idea, study design; D.B.—conception of idea, study design, interpretation of data, and reviewing of the manuscript. All authors have read and agreed to the published version of the manuscript.

Funding: This research was funded by the EU Regional Operational Program Calabria, Italy (POR Calabria FESR-FSE 2014–2020) Di, Me, Nu (prot. #52243/2017), and by the Department of Excellence (Italian Law. 232/2016) the Department of Pharmacy, and the Health and Nutritional Sciences, University of Calabria, Italy.

Institutional Review Board Statement: This study was conducted in accordance with the Declaration of Helsinki and approved by the Ethics Committee of the University of Calabria (#5727/2018).

Informed Consent Statement: Informed consent was obtained from all parents of adolescents involved in the study.

Data Availability Statement: The datasets used and/or analyzed in the current study are available from the corresponding author upon reasonable request.

Acknowledgments: We sincerely thank all participants enrolled from High School "Istituto Istruzione Superiore"—"Valentini-Majorana", Castrolibero (CS, Italy), and from sports teams of Cosenza and Paola, located in Calabria Region, Italy. We are also thankful to all partners of the Di.Me.Nu. project (Amphiios S.C.R.L., Istituto Istruzione Superiore—"Valentini-Majorana", Castrolibero, ELLE 17 S.R.L.).

Conflicts of Interest: The authors declare no conflict of interest.

References

1. Keys, A.; Menott, A.; Karvonen, M.J.; Aravanjs, C.; Blackburn, H.; Buzina, R.; Djordjevic, B.S.; Dontas, A.S.; Fidanza, F.; Keys, M.H.; et al. The Diet and 15-Year Death Rate in the Seven Countries Study. *Am. J. Epidemiol.* **2017**, *185*, 1130–1142. [CrossRef] [PubMed]
2. Martinez-Gonzalez, M.A.; Martin-Calvo, N. Mediterranean diet and life expectancy; beyond olive oil, fruits, and vegetables. *Curr. Opin. Clin. Nutr. Metab. Care* **2016**, *19*, 401–407. [CrossRef] [PubMed]
3. Widmer, R.J.; Flammer, A.J.; Lerman, L.O.; Lerman, A. The Mediterranean diet, its components, and cardiovascular disease. *Am. J. Med.* **2015**, *128*, 229–238. [CrossRef]
4. Mentella, M.C.; Scaldaferri, F.; Ricci, C.; Gasbarrini, A.; Miggiano, G.A.D. Cancer and Mediterranean Diet: A Review. *Nutrients* **2019**, *11*, 2059. [CrossRef]
5. Augimeri, G.; Montalto, F.I.; Giordano, C.; Barone, I.; Lanzino, M.; Catalano, S.; Ando, S.; De Amicis, F.; Bonofiglio, D. Nutraceuticals in the Mediterranean Diet: Potential Avenues for Breast Cancer Treatment. *Nutrients* **2021**, *13*, 2557. [CrossRef]
6. Lotfi, K.; Saneei, P.; Hajhashemy, Z.; Esmaillzadeh, A. Adherence to the Mediterranean Diet, Five-Year Weight Change, and Risk of Overweight and Obesity: A Systematic Review and Dose-Response Meta-Analysis of Prospective Cohort Studies. *Adv. Nutr.* **2022**, *13*, 152–166. [CrossRef] [PubMed]
7. Martin-Pelaez, S.; Fito, M.; Castaner, O. Mediterranean Diet Effects on Type 2 Diabetes Prevention, Disease Progression, and Related Mechanisms. A Review. *Nutrients* **2020**, *12*, 2236. [CrossRef] [PubMed]
8. Malmir, H.; Saneei, P.; Larijani, B.; Esmaillzadeh, A. Adherence to Mediterranean diet in relation to bone mineral density and risk of fracture: A systematic review and meta-analysis of observational studies. *Eur. J. Nutr.* **2018**, *57*, 2147–2160. [CrossRef]
9. Petersson, S.D.; Philippou, E. Mediterranean Diet, Cognitive Function, and Dementia: A Systematic Review of the Evidence. *Adv. Nutr.* **2016**, *7*, 889–904. [CrossRef] [PubMed]
10. Marventano, S.; Vetrani, C.; Vitale, M.; Godos, J.; Riccardi, G.; Grosso, G. Whole Grain Intake and Glycaemic Control in Healthy Subjects: A Systematic Review and Meta-Analysis of Randomized Controlled Trials. *Nutrients* **2017**, *9*, 769. [CrossRef] [PubMed]
11. Morrison, D.J.; Preston, T. Formation of short chain fatty acids by the gut microbiota and their impact on human metabolism. *Gut Microbes* **2016**, *7*, 189–200. [CrossRef]
12. Bullo, M.; Lamuela-Raventos, R.; Salas-Salvado, J. Mediterranean diet and oxidation: Nuts and olive oil as important sources of fat and antioxidants. *Curr. Top Med. Chem.* **2011**, *11*, 1797–1810. [CrossRef] [PubMed]
13. Jimenez-Lopez, C.; Carpena, M.; Lourenco-Lopes, C.; Gallardo-Gomez, M.; Lorenzo, J.M.; Barba, F.J.; Prieto, M.A.; Simal-Gandara, J. Bioactive Compounds and Quality of Extra Virgin Olive Oil. *Foods* **2020**, *9*, 1014. [CrossRef]
14. Bays, H.E.; Tighe, A.P.; Sadovsky, R.; Davidson, M.H. Prescription omega-3 fatty acids and their lipid effects: Physiologic mechanisms of action and clinical implications. *Expert Rev. Cardiovasc. Ther.* **2008**, *6*, 391–409. [CrossRef] [PubMed]
15. Santos-Buelga, C.; Gonzalez-Manzano, S.; Gonzalez-Paramas, A.M. Wine, Polyphenols, and Mediterranean Diets. What Else Is There to Say? *Molecules* **2021**, *26*, 5537. [CrossRef] [PubMed]
16. Subbaramaiah, K.; Michaluart, P.; Chung, W.J.; Tanabe, T.; Telang, N.; Dannenberg, A.J. Resveratrol inhibits cyclooxygenase-2 transcription in human mammary epithelial cells. *Ann. N. Y. Acad. Sci.* **1999**, *889*, 214–223. [CrossRef]
17. Vivancos, M.; Moreno, J.J. Effect of resveratrol, tyrosol and beta-sitosterol on oxidised low-density lipoprotein-stimulated oxidative stress, arachidonic acid release and prostaglandin E2 synthesis by RAW 264.7 macrophages. *Br. J. Nutr.* **2008**, *99*, 1199–1207. [CrossRef]
18. Altomare, R.; Cacciabaudo, F.; Damiano, G.; Palumbo, V.D.; Gioviale, M.C.; Bellavia, M.; Tomasello, G.; Lo Monte, A.I. The mediterranean diet: A history of health. *Iran. J. Public Health* **2013**, *42*, 449–457.
19. Bach-Faig, A.; Berry, E.M.; Lairon, D.; Reguant, J.; Trichopoulou, A.; Dernini, S.; Medina, F.X.; Battino, M.; Belahsen, R.; Miranda, G.; et al. Mediterranean diet pyramid today. Science and cultural updates. *Public Health Nutr.* **2011**, *14*, 2274–2284. [CrossRef]
20. Augimeri, G.; Bonofiglio, D. The Mediterranean Diet as a Source of Natural Compounds: Does It Represent a Protective Choice against Cancer? *Pharmaceuticals* **2021**, *14*, 920. [CrossRef]

21. Diolintzi, A.; Panagiotakos, D.B.; Sidossis, L.S. From Mediterranean diet to Mediterranean lifestyle: A narrative review. *Public Health Nutr.* **2019**, *22*, 2703–2713. [CrossRef] [PubMed]
22. Lake, A.A.; Mathers, J.C.; Rugg-Gunn, A.J.; Adamson, A.J. Longitudinal change in food habits between adolescence (11–12 years) and adulthood (32–33 years): The ASH30 Study. *J. Public Health* **2006**, *28*, 10–16. [CrossRef] [PubMed]
23. Strasser, B. Physical activity in obesity and metabolic syndrome. *Ann. N. Y. Acad. Sci.* **2013**, *1281*, 141–159. [CrossRef] [PubMed]
24. Booth, F.W.; Roberts, C.K.; Laye, M.J. Lack of exercise is a major cause of chronic diseases. *Compr. Physiol.* **2012**, *2*, 1143–1211. [CrossRef] [PubMed]
25. Morelli, C.; Avolio, E.; Galluccio, A.; Caparello, G.; Manes, E.; Ferraro, S.; Caruso, A.; De Rose, D.; Barone, I.; Adornetto, C.; et al. Nutrition Education Program and Physical Activity Improve the Adherence to the Mediterranean Diet: Impact on Inflammatory Biomarker Levels in Healthy Adolescents From the DIMENU Longitudinal Study. *Front. Nutr.* **2021**, *8*, 685247. [CrossRef]
26. Morelli, C.; Avolio, E.; Galluccio, A.; Caparello, G.; Manes, E.; Ferraro, S.; De Rose, D.; Santoro, M.; Barone, I.; Catalano, S.; et al. Impact of Vigorous-Intensity Physical Activity on Body Composition Parameters, Lipid Profile Markers, and Irisin Levels in Adolescents: A Cross-Sectional Study. *Nutrients* **2020**, *12*, 742. [CrossRef]
27. Augimeri, G.; Galluccio, A.; Caparello, G.; Avolio, E.; La Russa, D.; De Rose, D.; Morelli, C.; Barone, I.; Catalano, S.; Ando, S.; et al. Potential Antioxidant and Anti-Inflammatory Properties of Serum from Healthy Adolescents with Optimal Mediterranean Diet Adherence: Findings from DIMENU Cross-Sectional Study. *Antioxidants* **2021**, *10*, 1172. [CrossRef]
28. Galluccio, A.; Caparello, G.; Avolio, E.; Manes, E.; Ferraro, S.; Giordano, C.; Sisci, D.; Bonofiglio, D. Self-Perceived Physical Activity and Adherence to the Mediterranean Diet in Healthy Adolescents during COVID-19: Findings from the DIMENU Pilot Study. *Healthcare* **2021**, *9*, 622. [CrossRef]
29. World Health Organization. Global Action Plan on Physical Activity 2018–2030: More Active People for a Healthier World. Available online: https://apps.who.int/iris/handle/10665/272722. (accessed on 24 January 2022).
30. Serra-Majem, L.; Ribas, L.; Ngo, J.; Ortega, R.M.; Garcia, A.; Perez-Rodrigo, C.; Aranceta, J. Food, youth and the Mediterranean diet in Spain. Development of KIDMED, Mediterranean Diet Quality Index in children and adolescents. *Public Health Nutr.* **2004**, *7*, 931–935. [CrossRef]
31. World Health Organization. *Physical Status: The Use and Interpretation of Anthropometry. Report of A WHO Expert Committee*; World Health Organization: Geneva, Switzerland, 1995; Volume 854, pp. 1–452.
32. Società Italiana Di Nutrizione Umana. Available online: https://sinu.it/2019/07/09/fabbisogno-energetico-medioar-nellintervallo-deta-1-17-anni (accessed on 28 October 2020).
33. Malik, V.S.; Schulze, M.B.; Hu, F.B. Intake of sugar-sweetened beverages and weight gain: A systematic review. *Am. J. Clin. Nutr.* **2006**, *84*, 274–288. [CrossRef]
34. Neovius, M.; Linne, Y.; Rossner, S. BMI, waist-circumference and waist-hip-ratio as diagnostic tests for fatness in adolescents. *Int. J. Obes.* **2005**, *29*, 163–169. [CrossRef]
35. Velazquez-Lopez, L.; Santiago-Diaz, G.; Nava-Hernandez, J.; Munoz-Torres, A.V.; Medina-Bravo, P.; Torres-Tamayo, M. Mediterranean-style diet reduces metabolic syndrome components in obese children and adolescents with obesity. *BMC Pediatr.* **2014**, *14*, 175. [CrossRef]
36. Fletcher, E.A.; McNaughton, S.A.; Crawford, D.; Cleland, V.; Della Gatta, J.; Hatt, J.; Dollman, J.; Timperio, A. Associations between sedentary behaviours and dietary intakes among adolescents. *Public Health Nutr.* **2018**, *21*, 1115–1122. [CrossRef]
37. Bohara, S.S.; Thapa, K.; Bhatt, L.D.; Dhami, S.S.; Wagle, S. Determinants of Junk Food Consumption Among Adolescents in Pokhara Valley, Nepal. *Front. Nutr.* **2021**, *8*, 644650. [CrossRef] [PubMed]
38. Ajala, O.; English, P.; Pinkney, J. Systematic review and meta-analysis of different dietary approaches to the management of type 2 diabetes. *Am. J. Clin. Nutr.* **2013**, *97*, 505–516. [CrossRef]
39. Huo, R.; Du, T.; Xu, Y.; Xu, W.; Chen, X.; Sun, K.; Yu, X. Effects of Mediterranean-style diet on glycemic control, weight loss and cardiovascular risk factors among type 2 diabetes individuals: A meta-analysis. *Eur. J. Clin. Nutr.* **2015**, *69*, 1200–1208. [CrossRef] [PubMed]
40. Lasa, A.; Miranda, J.; Bullo, M.; Casas, R.; Salas-Salvado, J.; Larretxi, I.; Estruch, R.; Ruiz-Gutierrez, V.; Portillo, M.P. Comparative effect of two Mediterranean diets versus a low-fat diet on glycaemic control in individuals with type 2 diabetes. *Eur. J. Clin. Nutr.* **2014**, *68*, 767–772. [CrossRef] [PubMed]

MDPI

Article

Adherence to the Mediterranean Diet among Children and Youth in the Mediterranean Region in Croatia: A Comparative Study

Antonela Matana [1,*], Ivana Franić [1], Endica Radić Hozo [1], Ante Burger [1] and Petra Boljat [2]

[1] The University Department of Health Studies, University of Split, Ruđera Boškovića 35, 21000 Split, Croatia; ivana.zorica@ozs.unist.hr (I.F.); erhozo@ozs.unist.hr (E.R.H.); anteburger@gmail.com (A.B.)
[2] Elementary School Žnjan-Pazdigrad, Pazdigradska 1, 21000 Split, Croatia; petraa.boljat@gmail.com
* Correspondence: antonela.matana@gmail.com

Abstract: The Mediterranean diet (MD) is considered one of the healthiest dietary patterns. The aim of this study was to assess MD adherence in children and youth living in the Mediterranean region in Croatia and evaluate the differences in adherence to the MD among different educational stages. In total, 2722 individuals aged 2 to 24 years were enrolled in this study. Subjects were divided into different groups according to the Croatian educational system. Mediterranean Diet Quality Index (KIDMED) was used to assess adherence to the MD. In the total sample, the adherence to the MD was poor in 19.2%, average in 60.8%, and good in 20.1% of the study participants. The prevalence rate of poor adherence to the MD increased with higher educational stage, i.e., the highest prevalence rate of poor MD adherence was observed for college students (39.3%). Children having a higher number of snacks on days-off, those with lower physical activity, and not having breakfast together with a family are more likely to have poor MD adherence, while children having a higher number of snacks on working days are less likely to have a poor MD. The results of this study showed low adherence to the principles of the MD, confirming the need for improvement of adherence to the MD pattern in the studied population.

Keywords: Mediterranean diet; children; youth; kindergarten; primary school; secondary school; faculty

Citation: Matana, A.; Franić, I.; Radić Hozo, E.; Burger, A.; Boljat, P. Adherence to the Mediterranean Diet among Children and Youth in the Mediterranean Region in Croatia: A Comparative Study. *Nutrients* **2022**, *14*, 302. https://doi.org/10.3390/nu14020302

Academic Editor: Daniela Bonofiglio

Received: 24 December 2021
Accepted: 9 January 2022
Published: 12 January 2022

1. Introduction

The Mediterranean diet (MD) is considered one of the healthiest dietary patterns in the world, characterized by high consumption of fruits, vegetables, legumes, olive oil, nuts, and cereals, a moderate-high intake of fish, dairy products, and alcohol (mostly wine), and a low intake of saturated lipids, sweets, and red and processed meat [1,2]. Numerous studies have shown that adherence to the MD is associated with a significant reduction in total mortality and improvement in longevity, as well as with various health benefits, including prevention of cardiovascular diseases, neurodegenerative diseases, diabetes type 2, obesity, cancer, and many others [3–8].

However, despite all the existing evidence about the benefits of this diet, a transition from this dietary pattern towards a high-energy diet style, which is rich in saturated fats and low in micronutrients, has been observed, especially in the younger generation [9]. This change has led to an increase in obesity and numerous negative health-related consequences [6,10]. Childhood obesity is a particular public health concern [11]. The results of the European Childhood Obesity Surveillance Initiative in Croatia in 2018/2019 indicate that 35% of children aged 8.0 to 8.9 years had overweight or obesity [12]. In addition, according to the research conducted by the Organisation for Economic Cooperation and Development (OECD), in the next 30 years, life expectancy in Croatia will be shortened by 3.5 years due to overweight [13].

Healthy lifestyle habits develop in the early stages of life and impact human health significantly in later life, making childhood and adolescence particularly important for the adoption and maintenance of healthy habits [14]. So far, only a few studies on adherence to the MD among the Croatian young population have been performed [15–17]. A study recently conducted among Croatian university students showed that college students have poor eating habits, with 42.8% of students having low MD adherence scores [15]. Surprisingly, given the fact that kindergartens in Croatia follow institutionalized nutritional recommendations and that children spend most of their daytime hours in kindergartens, results for preschoolers in Croatia are inconsistent: while one study showed that only 6% of the children had a low MD adherence score (12), other recently published research revealed that almost half of the study participants (49%) had a low KIDMED index score [17].

To the best of our knowledge, there has been no study performed specifically among primary and secondary school children in Croatia, thus the question of what happens to children's eating habits after kindergarten age, especially considering that most of Croatian primary and secondary schools do not provide institutionalized feeding for their students, remains unanswered. Results of published studies suggest that good eating habits are lost by the time of university study, so it is necessary to determine at what age children's habits begin to change so that additional efforts can be made on time to educate children about the benefits of proper nutrition.

Given the aforesaid, estimating MD adherence and exploring potential predictors might be useful for developing strategies for improving diet quality. Therefore, the main aim of this study was to assess MD adherence in the youth population living in the Mediterranean region in Croatia and evaluate the differences in adherence to the MD and its components among preschool, primary, secondary school children and students by using the same validated questionnaire KIDMED for all age groups [18].

2. Materials and Methods

2.1. Study Design

This cross-sectional study was carried out from September to November 2021 in children and youths from the Mediterranean region of Croatia (including the regions of Istria, Kvarner, Dalmatia, the Dubrovnik area, and the Adriatic Islands). Participants were aged from 2 to 24 years and were enrolled in randomly selected public kindergartens, elementary or secondary schools, or faculties. The final sample comprised 2722 eligible participants. Participants were categorized into 5 groups according to the Croatian educational system: (i) kindergartens, (ii) primary schools (1st–4th grade), (iii) primary schools (5th–8th grade), (iv) secondary schools, and (v) faculties (college students). The study was approved by the Ethics Committee of the University Department of Health Studies, University of Split (Class 001-01/21-01/01, reg. no.: 2181-228-103/1-21-22) and was conducted in regulation with the latest Helsinki declaration. The subjects gave consent to participate by submitting a completed questionnaire.

2.2. Questionnaire

Data were collected using the anonymous questionnaire. Based on the study site preferences, the questionnaire was delivered either as a paper-based or online survey. The online survey was constructed with the Google Forms application and was distributed among study sites using email. The expected time to complete the survey was 10 min. For the purpose of this study, the questionnaire was completed by the child's parent for participants enrolled in kindergartens, primary and secondary schools, while university students filled out a questionnaire on their own. Only one child per household was included in the study.

The questionnaire consisted of four sections. In the first section, we collected general information about participants, including gender, age, type of study program, year of attendance and parent-reported (or self-reported for students) weight (in kg) and height (in cm). Additionally, the participant's general health (self- or parent-perceived) was rated as excellent, very good, good, or fair/poor. Body mass index (BMI) was calculated as weight divided by height squared (kg/m^2), in order to calculate the BMI-for-age percentiles using the CDC growth charts. According to the CDC classification, percentiles lower than 5th is considered as underweight, percentiles between 5th and 85th are considered as normal weight, percentiles between 85th and 95th are considered as overweight, and percentiles ≥95th are considered as obese. Specifically, according to the World Health Organization standards, students were considered to be underweight if their BMI was lower than 18.5, normal weight if the BMI was 18.5 to 24.9, overweight if the BMI was 25 to 29.9, and obese if it was greater than 30 [19].

The second section consisted of two questions regarding physical activity: "Do you/Does your child participate in organized physical activity (possible answers: Yes/No)?", "How many times a week do you (or does your child) do some sport, dance, or play a game in which you are (or your child is) very active? (possible answers: none, 1 time, 2–3 times, 4–5 times, 6 or more times)".

Dietary habits, such as the number of main meals and snacks (recorded separately for working days and off-days), as well as information on eating breakfast, lunch, and dinner together as a family, were assessed in the third section.

In the last section, the level of adherence to the MD for participants was evaluated using the KIDMED test (Mediterranean Diet Quality Index for children and adolescents) [18]. KIDMED is a questionnaire consisting of 16 yes or no questions. Questions with a negative connotation with a respect to MD were given a score of -1 (including consumption of fast food, baked goods, sweets, and skipping breakfast), and those with a positive connotation were given a score of +1 (consumption of oil, fish, fruits, vegetables, cereals, nuts, pulses, pasta or rice, dairy products, and yogurt). The total score ranges between −4 to 12 and is classified into 3 levels: (i) low MD adherence: KIDMED score ≤3; (ii) average MD adherence: KIDMED score 4–7; (iii) good MD adherence: KIDMED score ≥8. The instrument was originally developed to assess the level of adherence to the MD in Spanish children and adolescents aged 2 to 24, and was previously adapted for the Croatian language and tested for reliability and validity [15].

2.3. Statistical Analysis

The Kolmogorov-Smirnov test was used for normality checking. Due to the non-normal distribution of the data, continuous variables are presented as the median (interquartile range, IQR). Categorical variables are presented with frequencies (percentages). Differences in categorical variables were analyzed by using a Chi-square test, while the Kruskal-Wallis test was used for not normally distributed continuous variables.

Furthermore, we performed multivariate multinomial logistic regression in order to assess the association of MD adherence categories with the odds ratios of predictors that were significant in univariate models (including age, number of daily meals and snacks both on working days and days-off, two questions regarding physical activity, and having breakfast and dinner together as a family). BMI categories and having lunch together as a family were not significantly associated with MD adherence in the univariate model, therefore were not included in the final model.

Finally, a multivariable multinomial logistic regression was also employed to assess the simultaneous effect of MD adherence and level of physical activity on self-perceived health. p-values of less than 0.05 were considered statistically significant. Statistical analysis was conducted using Statistical Package Software for Social Science, version 28 (SPSS Inc., Chicago, IL, USA).

3. Results

A total of 2722 children and youths participated in this study. Basic characteristics of the study participants are presented in Table 1.

Table 1. Basic characteristics of the study participants.

Variable		Descriptive Statistics
Gender, *n* (%)		
	Females	1340 (49.2%)
	Males	1382 (50.8%)
Age, median (interquartile range)		10.0 (6.0)
BMI classification, *n* (%)		
	Underweight	199 (7.3%)
	Normal weight	1884 (69.2%)
	Overweight	351 (12.9%)
	Obese	163 (6%)
Educational stage, *n* (%)		
	Kindergarten	485 (17.8%)
	Primary school (1st–4th grade)	941 (34.5%)
	Primary school (5th–8th grade)	780 (28.6%)
	Secondary school	343 (12.6%)
	Faculty (college students)	173 (6.3%)

In the total sample, the median KIDMED index score was 6 (IQR: 3), while the adherence to the MD was poor in 19.2%, average in 60.7%, and good in 20.1% of the study participants. The highest compliance to the KIDMED items was observed for eating fast food less than once a week, the consumption of olive oil at home, eating breakfast, and the consumption of dairy products for breakfast (Table 2). No significant gender differences were observed for MD adherence categories ($p = 0.146$). Age was significantly associated with MD adherence categories (participants with poor MD adherence were the oldest, followed by the average group, while the youngest were participants from the good MD category ($p < 0.001$)). Although a higher prevalence of poor MD adherence was recorded among obese (25.6%) and overweight (20.8%) individuals compared to those with normal BMI (18.6%) or underweight (16.5%), no statistically significant association was observed for BMI categories and MD adherence ($p = 0.120$).

Regarding the differences in adherence to the MD for different educational stages, the results showed that the KIDMED index score decreased with higher educational stage, i.e., the highest KIDMED index score was observed for children enrolled in kindergartens, followed by children from the first four grades of primary schools, then children from grades 5–8 of primary schools and youths enrolled in secondary schools, while the lowest score was observed for students ($p < 0.001$) (Figure 1). The prevalence rate of poor adherence to MD also increased with higher educational stage (Table 2). The highest prevalence rate of poor MD adherence was observed for students (39.3%), then for children from secondary schools (25.7%), followed by primary school children (16.8% for 1st–4th grades and 19.6% for 5th–8th grades), and the lowest rate was observed for the children enrolled in kindergartens (11.3%) ($p < 0.001$) (Table 2).

Table 2. Results of the KIDMED test according to educational stage.

	Total Sample	Kindergarten	Primary School (1st–4th Grade)	Primary School (5th–8th Grade)	Secondary School	Faculty	*p* Value
KIDMED index score, *n* (%) [1]							
Poor	523 (19.2%)	55 (11.3%)	158 (16.8%)	153 (19.6%)	88 (25.7%)	69 (39.9%)	<0.001
Average	1653 (60.7%)	315 (64.9%)	599 (63.7%)	462 (59.2%)	196 (57.1%)	81 (46.8%)	
Good	546 (20.1%)	115 (23.7%)	184 (19.6%)	165 (21.2%)	59 (17.2%)	23 (13.3%)	
KIDMED items, *n* (%) [2]							
Fruit or fruit juice daily	2321 (85.3%)	455 (93.8%)	800 (85%)	652 (83.6%)	280 (81.6%)	129 (74.6%)	<0.001
Second serving of fruit daily	1271 (46.7%)	260 (53.6%)	437 (46.4%)	369 (47.3%)	144 (42.0%)	60 (34.7%)	<0.001
Fresh or cooked vegetables daily	1942 (71.4%)	376 (77.5%)	681 (72.4%)	538 (69%)	227 (66.2%)	117 (67.6%)	0.003
Fresh or cooked vegetables > 1/day	585 (21.5%)	107 (22.1%)	199 (21.1%)	172 (22.1%)	70 (20.4%)	37 (21.4%)	0.749
Regular fish consumption (at least 2–3/week)	658 (24.3%)	142 (29.3%)	230 (24.4%)	177 (22.7%)	77 (22.4%)	28 (16.2%)	0.010
	Total Sample	Kindergarten	Primary School (1st–4th Grade)	Primary School (5th–8th Grade)	Secondary School	Faculty	*p* Value
>1/week fast-food (hamburger) restaurant	130 (4.8%)	9 (1.9%)	17 (1.8%)	32 (4.1%)	41 (12%)	31 (17.9%)	<0.001
Pulses > 1/week	1556 (57.2%)	263 (54.2%)	545 (57.9%)	461 (59.1%)	194 (56.6%)	90 (52%)	0.429
Pasta or rice almost daily (≥5 days/week)	427 (15.7%)	49 (10.1%)	125 (13.3%)	113 (14.5%)	76 (22.2%)	62 (35.8%)	<0.001
Cereal or cereal product for breakfast	1555 (57.1%)	274 (56.5%)	559 (59.4%)	459 (58.8%)	200 (58.3%)	60 (34.7%)	<0.001
Regular nut consumption (at least 2–3/week)	1054 (38.7%)	172 (35.5%)	346 (36.8%)	305 (39.1%)	148 (43.1%)	81 (46.8%)	0.041
Use of olive oil at home	2486 (91.4%)	463 (95.5%)	862 (91.6%)	701 (89.9%)	306 (89.2%)	149 (86.1%)	0.003
No breakfast	299 (10.99%)	17 (3.5%)	45 (4.8%)	96 (12.3%)	73 (21.3%)	67 (38.7%)	<0.001
Dairy product for breakfast	2413 (88.7%)	446 (92%)	856 (91%)	700 (89.7%)	291 (84.8%)	116 (67.1%)	<0.001
Commercially baked goods or pastries for breakfast	1173 (43.1%)	167 (34.4%)	383 (40.7%)	361 (46.3%)	176 (51.3%)	84 (48.6%)	<0.001
Two yoghurts and/or 40 g cheese daily	1193 (43.84%)	239 (49.3%)	387 (41.1%)	343 (44%)	155 (45.2%)	68 (39.3%)	0.051
Sweets and candy several times a day	751 (27.6%)	134 (27.6%)	244 (25.9%)	226 (29%)	89 (25.9%)	58 (33.5%)	0.439

[1] Results of the KIDMED test as a categorical variable. [2] *n* (%) indicate the number of participants who answered affirmatively to each item. Statistically significant results are in bold.

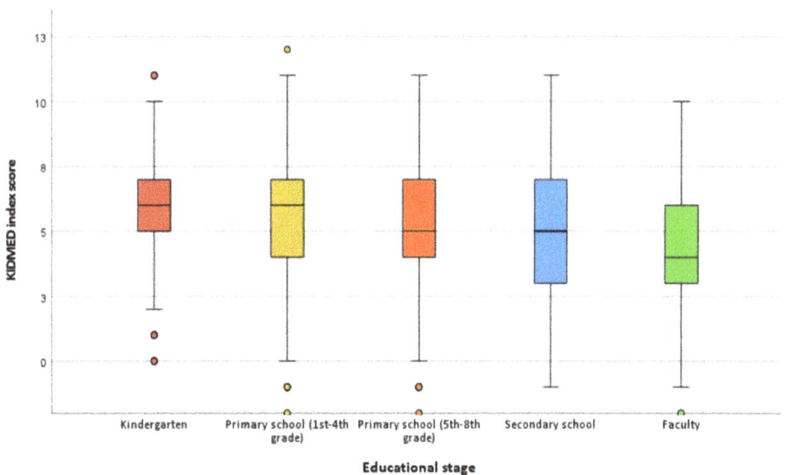

Figure 1. Box–plots for the KIDMED index score for different educational stages.

Basic descriptive statistics of the 16 KIDMED items according to educational stage is presented in Table 2. The prevalence rate of daily consumption of fruit or fruit juice, a second serving of fruit, fresh or cooked vegetables, as well as fish consumption more than 2–3 times a week, consumption of olive oil and dairy products for breakfast decreased with higher educational stage, while the prevalence rate of regular nut consumption and consumption of pasta and rice increased with higher educational stage. Children enrolled in kindergarten or first four grades of elementary school less frequently ate fast-foods and commercially baked goods or pastries for breakfast compared to higher educational stages. Students consumed cereals or cereal products for breakfast less frequently compared to individuals in other educational stages. The prevalence of skipping breakfast was higher for higher educational stages (Table 2).

Regarding physical activity, on the question "Do you/Does your child participate in organized physical activity?" the highest percentage of elementary school participants (78.7% for 1st–4th grades and 75.4% for 5th–8th grades, respectively) answered affirmatively compared to participants from high schools (47.5%), kindergartens (40.2%) and faculties (30.1%) ($p < 0.001$). Distribution of answers to the question "How many times a week do you (or does your child) do some sport, dance or play a game in which you are (or your child is) very active?" is shown in Figure 2. The "None" answer was more frequently chosen by students compared to other educational groups (Figure 2) ($p < 0.001$).

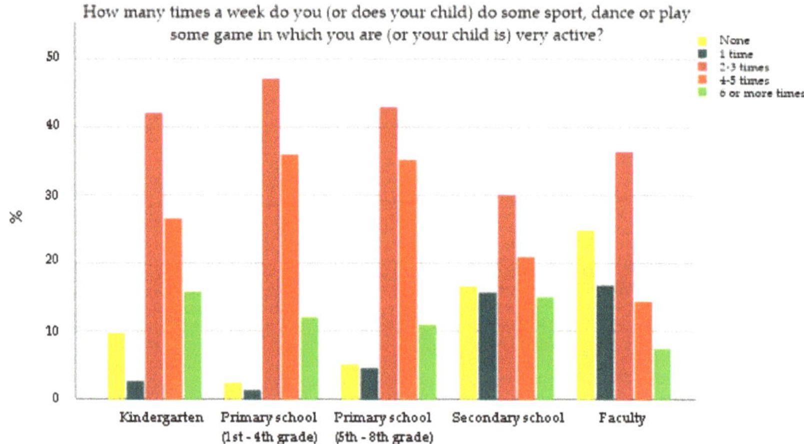

Figure 2. Distribution of answers to the question "How many times a week do you (or does your child) do some sport, dance or play some game in which you are (or your child is) very active?".

Furthermore, we performed a multivariate multinomial logistic regression for MD adherence with nine predictors listed in Table 3 ($p < 0.001$, Nagelkerke R2 = 0.103, correct prediction rate 61.9%) which were significant in the univariate model. The results of the multivariate model are presented in Table 3. Children having a higher number of snacks on days-off, those with lower physical activity, as assessed with the question: "How many times a week do you (or does your child) do some sport, dance or play some game in which you are (or your child is) very active?" and not having breakfast together with a family are more likely to have poor MD adherence, while children having a higher number of snacks on working days are less likely to have a poor MD adherence than average and good MD adherence (Table 3). Additionally, older participants were less likely to have average MD adherence than poor MD adherence.

Table 3. Results of the multinomial logistic regression with categories of MD adherence as a dependent variable.

Predictors	Average MD Adherence		Good MD Adherence	
	OR (95% CI) [1]	*p*-Value	OR (95% CI) [1]	*p*-Value
Age	0.962 (0.933, 0.991)	0.011	0.974 (0.939, 1.011)	0.165
Number of daily meals on working days	1.242 (0.913, 1.688)	0.167	1.341 (0.919, 1.957)	0.128
Number of daily meals on day-offs	1.264 (0.910, 1.757)	0.162	1.339 (0.903, 1.986)	0.146
Number of snacks on working days	1.470 (1.157, 1.867)	0.002	1.978 (1.470, 2.660)	<0.001
Number of snacks on day-offs	0.773 (0.627, 0.953)	0.016	0.738 (0.567, 0.960)	0.023
Do you/Does your child participate in organized physical activity?				
No	1.016 (0.776, 1.332)	0.907	0.819 (0.589, 1.139)	0.235
Predictors	Average MD Adherence		Good MD Adherence	
	OR (95% CI) [1]	*p*-Value	OR (95% CI) [1]	*p*-Value
Yes	-	-	-	-
How many times a week do you (or does your child) do some sport, dance or play some game in which you are (or your child is) very active?				
None	0.365 (0.216, 0.615)	<0.001	0.283 (0.145, 0.550)	<0.001
1 time	0.416 (0.236, 0.734)	0.002	0.374 (0.182, 0.770)	0.008
2–3 times	0.676 (0.463, 0.987)	0.043	0.508 (0.330, 0.782)	0.002
4–5 times	1.222 (0.814, 1.834)	0.334	0.759 (0.478, 1.206)	0.243
6 or more times	-	-	-	-
Having breakfast together as a family				
No	0.644 (0.514, 0.808)	<0.001	0.311 (0.233, 0.416)	<0.001
Yes	-	-	-	-
Having dinner together as a family				
No	0.980 (0.698, 1.375)	0.906	0.789 (0.493, 1.261)	0.322
Yes	-	-	-	-

[1] Odds ratios (OR) were calculated by multivariate multinomial logistic regression with low MD adherence as the reference category in the dependent variable.

The two-predictor model of participant's general health ($p < 0.001$) showed that both, lower level of physical activity and poor MD adherence, were associated with worse self-assessment of health, accounting for 6.3% of the total variance (Nagelkerke R^2), and the correct prediction rate was 58.8% (Table 4).

Table 4. Results of the multinomial logistic regression where participant's general health (self- or parent-perceived) was dependent variable while MD adherence and level of physical activity were independent variables.

Predictors	Good		Very Good	
	OR (95% CI)	*p*-Value	OR (95% CI)	*p*-Value
How many times a week do you (or does your child) do some sport, dance, or play some game in which you are (or your child is) very active?				
None	10.335 (3.140, 34.021)	<0.001	7.353 (2.582, 20.940)	<0.001
1 time	4.626 (2.565, 8.343)	<0.001	4.259 (2.759, 6.573)	<0.001
2–3 times	1.626 (0.978, 2.705)	0.061	2.804 (1.990, 3.950)	<0.001
4–5 times	0.886 (0.515, 1.525)	0.662	2.196 (1.552, 3.108)	<0.001
6 or more times	-	-	-	-
Mediterranean index score classification				
Average MD adherence	0.596 (0.419, 0.848)	0.004	0.812 (0.643, 1.026)	0.081
Good MD adherence	0.497 (0.311, 0.794)	0.003	0.646 (0.485, 0.861)	0.003
Poor MD adherence	-	-	-	-

Odds ratios (OR) were calculated by multivariate multinomial logistic regression with excellent general health (self- or parent-perceived) as the reference category in the dependent variable. Participants' general health was rated as excellent, very good, good, or fair/poor; however, no "fair/poor" rating was recorded in our sample.

4. Discussion

The main purpose of this study was to evaluate the differences in adherence to the Mediterranean diet and its components according to different educational stages in children and youths living in the Mediterranean region in Croatia. To the best of our knowledge, this is the first study that evaluates and compares MD adherence in individuals from all educational stages, including kindergartens, primary and secondary schools, and faculties.

Our results demonstrated a rather low prevalence of good adherence to the MD over the entire sample (20.1%). This result is in line with a study performed in the Croatian adult population, in which only 23% of the participants from Southern Dalmatia adhered to the principles of MD [20]. Furthermore, we showed that the prevalence rate of poor adherence to the MD increased with higher education stage. The highest prevalence rate of poor MD adherence was recorded among students (39.3%), then for children from secondary schools (25.7%), followed by primary school children (19.6% for 1st–4th grades and 16.8% for 5th–8th grades), while the lowest rate was observed for the children enrolled in kindergartens (11.3%). These results were somehow expected, given the fact that kindergartens in Croatia follow institutionalized nutritional recommendations which promote increased consumption of vegetables, fruits, meat, fish, and dairy products, while the great majority of primary and high schools do not provide institutionalized feeding for their students [16]. Besides the probable influence of institutionalized feeding, this result could also be explained by parental supervision and control over children's diets, an influence that is gradually lost as children grow up. Several studies have shown that parental control is associated with following healthy dietary habits [21–23]. Additionally, another potential factor associated with food choice is pocket money amount which usually increases with age, and increases the probability of consumption of unhealthy fast-food, baked goods, and sweets [24]. Moreover, it is known that students are more likely to buy foods that are fast, convenient, and inexpensive [25]. In the study performed by Marquis et al., it was also shown that college students often make food choices based on cost and convenience over health [26]. Our finding that the prevalence rate of poor adherence to the MD increased with higher education stage is in line with other similar studies performed in Spain and Italy [18,24,27,28]. Croatian data on children and young individuals are relatively few and mainly refer to preschool children and students. We performed the largest study of MD adherence in the Croatian youth population so far by including 2722 subjects. Our results for students are in accordance with the study from Štefan et al., in which 42.8% of college students had poor compliance with MD [15]. However, as already mentioned, results for the preschool population are contradictory, while the results of one study showed that only 6% of children had low KIDMED score [16], the other identifier of significantly higher prevalence (49%) of low adherence to the MD [17]. In both studies, only children from the urban area of Split-Dalmatia County were examined, while our study included subjects from the entire Mediterranean region of Croatia. To the best of our knowledge, there is no published study performed among primary and secondary school children in Croatia, so the results of this study are of particular interest for this young population.

A statistically significant difference in the KIDMED score with regard to sex and BMI categories was not found in our study, which is in line with the results of the majority of other studies, including the systematic reviews of European data [17,29–32].

Moreover, the multinomial logistic regression results identified several predictors of MD adherence. Children not having breakfast together with a family, having a higher number of snacks on days-off and a lower number of snacks on working days, and those with lower physical activity are more likely to have poor MD adherence.

Several studies have shown that eating at least one meal per day with a family member has a positive impact on general health and avoiding obesity [33]. As in the present study, studies carried out in Italy and Spain also observed a significant association between MD adherence and eating breakfast with the family [34,35].

Interestingly, our results indicate that children having a higher number of snacks on days-off and a lower number of snacks on working days are more likely to have poor MD adherence. Actually, several studies conducted in the Nordic countries and in the United States have observed differences in dietary quality on working days compared to the weekend [36–41]. The study in Swedish children has shown that children had their highest intake of sucrose on Fridays and Saturdays due to increased intake of sweets and soft drinks [38]. Another study also confirmed that the intake of total sugars and foods and drinks rich in added sugar were generally higher on weekends versus weekdays for children in Hungary, Italy, and Sweden [36]. Furthermore, during days off, children have more free time and are likely to spend more time in front of screens (watching television, using computers, tablets, and smartphones, and playing video games) [42]. Screen time is usually associated with sedentary behavior and snacking which is often characterized by low nutritional quality [43]. Moreover, another study has shown that a greater amount of screen time is associated with lower consumption of healthy food, including vegetables, legumes, fish, and nuts, and greater consumption of sweets and fast food which consequently leads to poor adherence to the principles of MD [44].

Regarding physical activity, our results are in accordance with previous reports, showing a positive association between physical activity and other healthy lifestyle habits including proper nutrition and MD adherence [24,31,45–49]. A possible explanation for this association is that those children who are physically active and eat healthily are probably adequately educated and coached by their parents. Indeed, previous reports of a positive association between general parental support and physical activity among youth have been recorded [50–52]. Results from an above mentioned study performed in Croatian preschoolers also confirmed a positive association of physical activity and KIDMED scores [16], while other Croatian studies performed in youths did not examine physical activity [15,17]. Furthermore, the results of the present study confirmed that a combination of adherence to MD and high physical activity is beneficial to participants' parent- or self-rated health, and underscore the importance of adopting healthy lifestyle habits for better general health. Although overall health status in the present study was self- or parent-reported, it has been shown that self-perceived health is a valid proxy indicator of health status [53] and that both child and parent reports for health-related quality of life are valid [54].

Our study has several limitations that need to be mentioned. The main limitation is the cross-sectional design, which limits inference on causality. Second, the parents filled out the questionnaire (for all study subjects except for the students), so there is a possibility of parental overestimation or underestimation of the quality of a child's nutrition. Data on some other potential predictors of adherence to an MD were not collected, such as sleep habits and socio-economic data. On the other hand, this is the most comprehensive study conducted so far on MD adherence in the youth population in Croatia, the results of which have undoubtedly improved existing knowledge on adherence to an MD among children and youth in the Mediterranean region of Croatia. We have included subjects from a larger geographical region and within a wider age range compared to previously published studies in preschool children and students. Even more, to the best of our knowledge, we performed the first study on primary and secondary school children from Croatia and provided crucial data on dietary habits for this population. Another strength of this study is that we used the same validated questionnaire KIDMED for all educational stages, including university students, which enables us to directly compare the results among groups.

5. Conclusions

To conclude, the results of this study showed low adherence to the principles of the MD, confirming the need for the improvement of adherence to the MD pattern in the studied population. The present study provided crucial information on MD adherence in different educational stages and helped in defining periods when dietary habits are less healthy. This can help in developing tailored nutritional programs since it was shown that strategies designed explicitly to subgroups are needed. The findings of this study underscore the need to advise and motivate the young population so that healthy dietary habits can be integrated into their lifestyle.

Author Contributions: Conceptualization, A.M.; methodology, A.M.; formal analysis, A.M.; investigation, A.M., I.F., E.R.H., A.B., P.B.; resources, A.M., I.F., E.R.H., A.B., P.B.; data curation, A.M.; writing—original draft preparation, A.M.; writing—review and editing, A.M., I.F., E.R.H., A.B., P.B.; visualization, A.M.; supervision, A.M. All authors have read and agreed to the published version of the manuscript.

Funding: This research was funded by the institutional project "Adherence to the pattern of the Mediterranean diet and the level of physical activity in children and youth in Croatia" (SOZS-IP-2021-1) from the University Department of Health Studies, University of Split, Split, Croatia.

Institutional Review Board Statement: The study was conducted in accordance with the Declaration of Helsinki, and approved by the Ethics Committee of the University Department of Health Studies, University of Split (Class 001-01/21-01/01, reg. no.: 2181-228-103/1-21-22).

Informed Consent Statement: Informed consent was obtained from all parents of children, as well as from students, involved in the study. The subjects gave consent to participate by submitting a completed questionnaire.

Data Availability Statement: Raw data can be found at corresponding author via e-mail: antonela.matana@gmail.com.

Conflicts of Interest: The authors declare no conflict of interest.

References

1. Marques, G.F.S.; Pinto, S.M.O.; Reis, A.; Martins, T.D.B.; Conceicao, A.P.D.; Pinheiro, A.R.V. Adherence to the Mediterranean Diet in Elementary School Children (1st Cycle). *Rev. Paul. Pediatr.* **2021**, *39*, e2019259. [CrossRef]
2. Grosso, G.; Buscemi, S.; Galvano, F.; Mistretta, A.; Marventano, S.; La Vela, V.; Drago, F.; Gangi, S.; Basile, F.; Biondi, A. Mediterranean diet and cancer: Epidemiological evidence and mechanism of selected aspects. *BMC Surg.* **2013**, *13* (Suppl. 2), S14. [CrossRef] [PubMed]

3. Esposito, K.; Maiorino, M.I.; Bellastella, G.; Chiodini, P.; Panagiotakos, D.; Giugliano, D. A journey into a Mediterranean diet and type 2 diabetes: A systematic review with meta-analyses. *BMJ Open* **2015**, *5*, e008222. [CrossRef] [PubMed]
4. Mattioli, A.V.; Palmiero, P.; Manfrini, O.; Puddu, P.E.; Nodari, S.; Dei Cas, A.; Mercuro, G.; Scrutinio, D.; Palermo, P.; Sciomer, S.; et al. Mediterranean diet impact on cardiovascular diseases: A narrative review. *J. Cardiovasc. Med.* **2017**, *18*, 925–935. [CrossRef] [PubMed]
5. Schwingshackl, L.; Schwedhelm, C.; Galbete, C.; Hoffmann, G. Adherence to Mediterranean Diet and Risk of Cancer: An Updated Systematic Review and Meta-Analysis. *Nutrients* **2017**, *9*, 1063. [CrossRef]
6. Grosso, G.; Marventano, S.; Yang, J.; Micek, A.; Pajak, A.; Scalfi, L.; Galvano, F.; Kales, S.N. A comprehensive meta-analysis on evidence of Mediterranean diet and cardiovascular disease: Are individual components equal? *Crit. Rev. Food Sci. Nutr.* **2017**, *57*, 3218–3232. [CrossRef]
7. Vidal-Peracho, C.; Tricas-Moreno, J.M.; Lucha-Lopez, A.C.; Lucha-Lopez, M.O.; Camunas-Pescador, A.C.; Caverni-Munoz, A.; Fanlo-Mazas, P. Adherence to Mediterranean Diet Pattern among Spanish Adults Attending a Medical Centre: Nondiabetic Subjects and Type 1 and 2 Diabetic Patients. *J. Diabetes Res.* **2017**, *2017*, 5957821. [CrossRef]
8. Liyanage, T.; Ninomiya, T.; Wang, A.; Neal, B.; Jun, M.; Wong, M.G.; Jardine, M.; Hillis, G.S.; Perkovic, V. Effects of the Mediterranean Diet on Cardiovascular Outcomes-A Systematic Review and Meta-Analysis. *PLoS ONE* **2016**, *11*, e0159252. [CrossRef]
9. Tsakiraki, M.; Grammatikopoulou, M.G.; Stylianou, C.; Tsigga, M. Nutrition transition and health status of Cretan women: Evidence from two generations. *Public Health Nutr.* **2011**, *14*, 793–800. [CrossRef]
10. Belahsen, R. Nutrition transition and food sustainability. *Proc. Nutr. Soc.* **2014**, *73*, 385–388. [CrossRef]
11. Karnik, S.; Kanekar, A. Childhood obesity: A global public health crisis. *Int. J. Prev. Med.* **2012**, *3*, 1–7. [PubMed]
12. Musić Milanović, S.; Lang Morović, M.; Križan, H. *WHO European Childhood Obesity Surveillance Initiative (COSI), Croatia 2018/2019 (CroCOSI)*; Croatian Institute of Public Health: Zagreb, Croatia, 2021.
13. OECD. *Heavy Burden of Obesity: The Economics of Prevention—A Quick Guide for Policy Makers*; OECD: Paris, France, 2019.
14. De Santi, M.; Callari, F.; Brandi, G.; Toscano, R.V.; Scarlata, L.; Amagliani, G.; Schiavano, G.F. Mediterranean diet adherence and weight status among Sicilian Middle school adolescents. *Int. J. Food Sci. Nutr.* **2020**, *71*, 1010–1018. [CrossRef]
15. Stefan, L.; Prosoli, R.; Juranko, D.; Cule, M.; Milinovic, I.; Novak, D.; Sporis, G. The Reliability of the Mediterranean Diet Quality Index (KIDMED) Questionnaire. *Nutrients* **2017**, *9*, 419. [CrossRef] [PubMed]
16. Salcin, L.O.; Karin, Z.; Miljanovic Damjanovic, V.; Ostojic, M.; Vrdoljak, A.; Gilic, B.; Sekulic, D.; Lang-Morovic, M.; Markic, J.; Sajber, D. Physical Activity, Body Mass, and Adherence to the Mediterranean Diet in Preschool Children: A Cross-Sectional Analysis in the Split-Dalmatia County (Croatia). *Int. J. Environ. Res. Public Health* **2019**, *16*, 3237.
17. Nenadić, D.B.; Kolak, E.; Selak, M.; Smoljo, M.; Radić, J.; Vučković, M.; Dropuljić, B.; Pijerov, T.; Cikoš, D.B. Anthropometric Parameters and Mediterranean Diet Adherence in Preschool Children in Split-Dalmatia County, Croatia—Are They Related? *Nutrients* **2021**, *13*, 4252.
18. Serra-Majem, L.; Ribas, L.; Ngo, J.; Ortega, R.M.; García, A.; Pérez-Rodrigo, C.; Aranceta, J. Food, youth and the Mediterranean diet in Spain. Development of KIDMED, Mediterranean Diet Quality Index in children and adolescents. *Public Health Nutr.* **2004**, *7*, 931–935. [CrossRef]
19. World Health Organization. *Obesity: Preventing and Managing the Global Epidemic. Report of a WHO Consultation*; WHO Technical Report Series 894; World Health Organization: Geneva, Switzerland, 2000; pp. 1–253.
20. Kolcic, I.; Relja, A.; Gelemanovic, A.; Miljkovic, A.; Boban, K.; Hayward, C.; Rudan, I.; Polasek, O. Mediterranean diet in the southern Croatia—does it still exist? *Croat. Med. J.* **2016**, *57*, 415–424. [CrossRef]
21. Shier, V.; Nicosia, N.; Datar, A. Neighborhood and home food environment and children's diet and obesity: Evidence from military personnel's installation assignment. *Soc. Sci. Med.* **2016**, *158*, 122–131. [CrossRef]
22. Neumark-Sztainer, D.; Hannan, P.J.; Story, M.; Croll, J.; Perry, C. Family meal patterns: Associations with sociodemographic characteristics and improved dietary intake among adolescents. *J. Am. Diet. Assoc.* **2003**, *103*, 317–322. [CrossRef]
23. Larson, N.I.; Neumark-Sztainer, D.; Hannan, P.J.; Story, M. Family meals during adolescence are associated with higher diet quality and healthful meal patterns during young adulthood. *J. Am. Diet. Assoc.* **2007**, *107*, 1502–1510. [CrossRef]
24. Arcila-Agudelo, A.M.; Ferrer-Svoboda, C.; Torres-Fernandez, T.; Farran-Codina, A. Determinants of Adherence to Healthy Eating Patterns in a Population of Children and Adolescents: Evidence on the Mediterranean Diet in the City of Mataro (Catalonia, Spain). *Nutrients* **2019**, *11*, 854. [CrossRef]
25. Deliens, T.; Clarys, P.; De Bourdeaudhuij, I.; Deforche, B. Determinants of eating behaviour in university students: A qualitative study using focus group discussions. *BMC Public Health* **2014**, *14*, 53. [CrossRef] [PubMed]
26. Marquis, M. Exploring convenience orientation as a food motivation for college students living in residence halls. *Int. J. Consum. Stud.* **2005**, *29*, 55–63. [CrossRef]
27. Grosso, G.; Marventano, S.; Buscemi, S.; Scuderi, A.; Matalone, M.; Platania, A.; Giorgianni, G.; Rametta, S.; Nolfo, F.; Galvano, F.; et al. Factors Associated with Adherence to the Mediterranean Diet among Adolescents Living in Sicily, Southern Italy. *Nutrients* **2013**, *5*, 4908–4923. [CrossRef]
28. Mariscal-Arcas, M.; Rivas, A.; Velasco, J.; Ortega, M.; Caballero, A.M.; Olea-Serrano, F. Evaluation of the Mediterranean Diet Quality Index (KIDMED) in children and adolescents in Southern Spain. *Public Health Nutr.* **2009**, *12*, 1408–1412. [CrossRef]

29. Sahingoz, S.A.; Sanlier, N. Compliance with Mediterranean Diet Quality Index (KIDMED) and nutrition knowledge levels in adolescents. A case study from Turkey. *Appetite* **2011**, *57*, 272–277. [CrossRef] [PubMed]
30. Cabrera, S.G.; Fernandez, N.H.; Hernandez, C.R.; Nissensohn, M.; Roman-Vinas, B.; Serra-Majem, L. KIDMED test; prevalence of low adherence to the Mediterranean Diet in children and young; a systematic review. *Nutr. Hosp.* **2015**, *32*, 2390–2399.
31. Iaccarino Idelson, P.; Scalfi, L.; Valerio, G. Adherence to the Mediterranean Diet in children and adolescents: A systematic review. *Nutr. Metab. Cardiovasc. Dis.* **2017**, *27*, 283–299. [CrossRef]
32. Agostinis-Sobrinho, C.; Santos, R.; Rosario, R.; Moreira, C.; Lopes, L.; Mota, J.; Martinkenas, A.; Garcia-Hermoso, A.; Correa-Bautista, J.E.; Ramirez-Velez, R. Optimal Adherence to a Mediterranean Diet May Not Overcome the Deleterious Effects of Low Physical Fitness on Cardiovascular Disease Risk in Adolescents: A Cross-Sectional Pooled Analysis. *Nutrients* **2018**, *10*, 815. [CrossRef]
33. Rollins, B.Y.; Belue, R.Z.; Francis, L.A. The beneficial effect of family meals on obesity differs by race, sex, and household education: The national survey of children's health, 2003–2004. *J. Am. Diet. Assoc.* **2010**, *110*, 1335–1339. [CrossRef]
34. Roccaldo, R.; Censi, L.; D'Addezio, L.; Toti, E.; Martone, D.; D'Addesa, D.; Cernigliaro, A.; ZOOM8 Study group. Adherence to the Mediterranean diet in Italian school children (The ZOOM8 Study). *Int. J. Food Sci. Nutr.* **2014**, *65*, 621–628. [CrossRef] [PubMed]
35. Shalá, A.; López-Guimera, G.; Fauquet, J.; Puntí, J.; Leiva, D.; Sánchez-Carracedo, D. Association between Family Meals and the Adherence to the Mediterranean Diet in Spanish Adolescents. *J. Child Adolesc. Behav.* **2017**, *5*, 272–366. [CrossRef]
36. Svensson, A.; Larsson, C.; Eiben, G.; Lanfer, A.; Pala, V.; Hebestreit, A.; Huybrechts, I.; Fernandez-Alvira, J.M.; Russo, P.; Koni, A.C.; et al. European children's sugar intake on weekdays versus weekends: The IDEFICS study. *Eur. J. Clin. Nutr.* **2014**, *68*, 822–828. [CrossRef] [PubMed]
37. Lehtisalo, J.; Erkkola, M.; Tapanainen, H.; Kronberg-Kippila, C.; Veijola, R.; Knip, M.; Virtanen, S.M. Food consumption and nutrient intake in day care and at home in 3-year-old Finnish children. *Public Health Nutr.* **2010**, *13*, 957–964. [CrossRef]
38. Garemo, M.; Lenner, R.A.; Strandvik, B. Swedish pre-school children eat too much junk food and sucrose. *Acta Paediatr.* **2007**, *96*, 266–272. [CrossRef] [PubMed]
39. Haines, P.S.; Hama, M.Y.; Guilkey, D.K.; Popkin, B.M. Weekend eating in the United States is linked with greater energy, fat, and alcohol intake. *Obes. Res.* **2003**, *11*, 945–949. [CrossRef] [PubMed]
40. Sepp, H.; Lennernas, M.; Pettersson, R.; Abrahamsson, L. Children's nutrient intake at preschool and at home. *Acta Paediatr.* **2001**, *90*, 483–491. [CrossRef]
41. Hart, C.N.; Raynor, H.A.; Osterholt, K.M.; Jelalian, E.; Wing, R.R. Eating and activity habits of overweight children on weekdays and weekends. *Int. J. Pediatr. Obes.* **2011**, *6*, 467–472. [CrossRef] [PubMed]
42. Jago, R.; Thompson, J.L.; Sebire, S.J.; Wood, L.; Pool, L.; Zahra, J.; Lawlor, D.A. Cross-sectional associations between the screen-time of parents and young children: Differences by parent and child gender and day of the week. *Int. J. Behav. Nutr. Phys. Act.* **2014**, *11*, 54. [CrossRef] [PubMed]
43. Feeley, A.; Musenge, E.; Pettifor, J.M.; Norris, S.A. Changes in dietary habits and eating practices in adolescents living in urban South Africa: The birth to twenty cohort. *Nutrition* **2012**, *28*, e1–e6. [CrossRef] [PubMed]
44. Warnberg, J.; Perez-Farinos, N.; Benavente-Marin, J.C.; Gomez, S.F.; Labayen, I.; AGZ; Gusi, N.; Aznar, S.; Alcaraz, P.E.; Gonzalez-Valeiro, M.; et al. Screen Time and Parents' Education Level Are Associated with Poor Adherence to the Mediterranean Diet in Spanish Children and Adolescents: The PASOS Study. *J. Clin. Med.* **2021**, *10*, 795. [CrossRef]
45. Llargues, E.; Franco, R.; Recasens, A.; Nadal, A.; Vila, M.; Perez, M.J.; Manresa, J.M.; Recasens, I.; Salvador, G.; Serra, J.; et al. Assessment of a school-based intervention in eating habits and physical activity in school children: The AVall study. *J. Epidemiol. Community Health* **2011**, *65*, 896–901. [CrossRef] [PubMed]
46. Bibiloni, M.d.M.; Pich, J.; Cordova, A.; Pons, A.; Tur, J.A. Association between sedentary behaviour and socioeconomic factors, diet and lifestyle among the Balearic Islands adolescents. *BMC Public Health* **2012**, *12*, 718. [CrossRef] [PubMed]
47. Martinez-Gomez, D.; Veiga, O.L.; Gomez-Martinez, S.; Zapatera, B.; Calle, M.E.; Marcos, A.; AFINOS Study Group. Behavioural correlates of active commuting to school in Spanish adolescents: The AFINOS (Physical Activity as a Preventive Measure Against Overweight, Obesity, Infections, Allergies, and Cardiovascular Disease Risk Factors in Adolescents) study. *Public Health Nutr.* **2011**, *14*, 1779–1786. [CrossRef]
48. Moliner-Urdiales, D.; Ruiz, J.R.; Ortega, F.B.; Rey-Lopez, J.P.; Vicente-Rodriguez, G.; Espana-Romero, V.; Munguia-Izquierdo, D.; Castillo, M.J.; Sjostrom, M.; Moreno, L.A.; et al. Association of objectively assessed physical activity with total and central body fat in Spanish adolescents; The HELENA Study. *Int. J. Obes.* **2009**, *33*, 1126–1135. [CrossRef] [PubMed]
49. Wittmeier, K.D.M.; Mollard, R.C.; Kriellaars, D.J. Physical activity intensity and risk of overweight and adiposity in children. *Obesity* **2008**, *16*, 415–420. [CrossRef] [PubMed]
50. Beets, M.W.; Cardinal, B.J.; Alderman, B.L. Parental social support and the physical activity-related behaviors of youth: A review. *Health Educ. Behav.* **2010**, *37*, 621–644. [CrossRef]
51. Yao, C.A.; Rhodes, R.E. Parental correlates in child and adolescent physical activity: A meta-analysis. *Int. J. Behav. Nutr. Phys. Act.* **2015**, *12*, 10. [CrossRef]
52. Edwardson, C.L.; Gorely, T. Parental influences on different types and intensities of physical activity in youth: A systematic review. *Psychol. Sport Exerc.* **2010**, *11*, 522–535. [CrossRef]

53. Martikainen, P.; Aromaa, A.; Heliovaara, M.; Klaukka, T.; Knekt, P.; Maatela, J.; Lahelma, E. Reliability of perceived health by sex and age. *Soc. Sci.Med.* **1999**, *48*, 1117–1122. [CrossRef]
54. Theunissen, N.C.; Vogels, T.G.; Koopman, H.M.; Verrips, G.H.; Zwinderman, K.A.; Verloove-Vanhorick, S.P.; Wit, J.M. The proxy problem: Child report versus parent report in health-related quality of life research. *Qual. Life Res.* **1998**, *7*, 387–397. [CrossRef] [PubMed]

Article

Adherence to the Mediterranean Diet and Its Association with the Level of Physical Activity in Fitness Center Users: Croatian-Based Study

Dinko Martinovic [1], Daria Tokic [2], Lovre Martinovic [1], Marko Kumric [1], Marino Vilovic [1], Doris Rusic [3], Josip Vrdoljak [1], Ivan Males [4], Tina Ticinovic Kurir [1], Slaven Lupi-Ferandin [5] and Josko Bozic [1,*]

1 Department of Pathophysiology, University of Split School of Medicine, 21000 Split, Croatia; dinko.martinovic@mefst.hr (D.M.); lm71805@mefst.hr (L.M.); marko.kumric@mefst.hr (M.K.); marino.vilovic@mefst.hr (M.V.); josip.vrdoljak@mefst.hr (J.V.); tticinov@mefst.hr (T.T.K.)
2 Department of Anesthesiology and Intensive Care, University Hospital of Split, 21000 Split, Croatia; dtokic@kbsplit.hr
3 Department of Pharmacy, University of Split School of Medicine, 21000 Split, Croatia; doris.rusic@mefst.hr
4 Department of Surgery, University Hospital of Split, 21000 Split, Croatia; ivmales@kbsplit.hr
5 Department of Maxillofacial Surgery, University Hospital of Split, 21000 Split, Croatia; slupif@kbsplit.hr
* Correspondence: josko.bozic@mefst.hr; Tel.: +385-21-557-871

Abstract: The Mediterranean diet (MD) is based on the traditional cuisine of south European countries, and it is considered one of the healthiest dietary patterns worldwide. The promotion of combined MD and physical activity has shown major benefits. However, the association between physical activity and the MD in regular fitness center users is still insufficiently investigated. This cross-sectional survey-based study was conducted on 1220 fitness center users in Croatia. The survey consisted of three parts: general information, the Mediterranean Diet Serving Score (MDSS) and the International Physical Activity Questionnaire Short Form (IPAQ-SF). The results showed that 18.6% of fitness center users were adherent to the MD, and there was a significant positive correlation between the level of physical activity and the MDSS score (r = 0.302, $p < 0.001$). Moreover, after dividing the sample into tertiles based on the IPAQ-SF score, the third tertile (MET > 3150 min/wk) had the most fitness center users (34.4%) adherent to the MD, while the first tertile (MET < 1750 min/wk) had the least (6.1%). These outcomes emphasize the importance of physical activity as they imply that, with higher levels of physical activity, people are also possibly more aware of the importance that a healthy and balanced diet has on their well-being.

Keywords: Mediterranean diet; physical activity; dietary supplements; fitness; MDSS; IPAQ-SF

Citation: Martinovic, D.; Tokic, D.; Martinovic, L.; Kumric, M.; Vilovic, M.; Rusic, D.; Vrdoljak, J.; Males, I.; Ticinovic Kurir, T.; Lupi-Ferandin, S.; et al. Adherence to the Mediterranean Diet and Its Association with the Level of Physical Activity in Fitness Center Users: Croatian-Based Study. *Nutrients* **2021**, *13*, 4038. https://doi.org/10.3390/nu13114038

Academic Editor: Daniela Bonofiglio

Received: 23 October 2021
Accepted: 9 November 2021
Published: 12 November 2021

1. Introduction

It is well established that regular exercise combined with a diversified and balanced diet can lead to a longer, healthy and more satisfying life [1–3]. Physical activity as a beneficial mechanism on health was first investigated in 1950s in correlation with cardiovascular diseases [4]. Since then, exercise was thoroughly researched for its beneficial impact on health. In addition to reducing the risk of developing cardiovascular diseases, it was established that physical activity also has a positive impact on preventing serious illnesses such as diabetes, depression and obesity [5,6]. As aforementioned, a healthy diet is as important as exercise, and an adequate intake of macronutrients and micronutrients with proper hydration is the base of a balanced diet. Furthermore, besides a balanced diet, physically active people also commonly use dietary supplements to enhance results and ensure optimal nutrient intake [7]. Dietary supplements can be defined as food additions containing higher levels of proteins, vitamins and other micronutrients made to amplify and boost regular diet [8].

Whereas some of the widespread diets have been developed in cooperation with nutritionists and physicians, some diets are based on the tradition of a specific region or a country [9]. The Mediterranean diet (MD) is based on the traditional cuisine of south European countries on the shores of the Mediterranean Sea. The foundation of this diet is the high intake of olive oil, vegetables, fruits, nuts, cereals and legumes, while the intake of fish, red wine, dairy products and meat should be moderate [10]. However, white meat should be consumed more frequently than red meat. Moreover, individuals should maintain a small, continuous consumption of red wine, and the recommendation for men and women is 1–2 glasses/day and 1 glass/day, respectively. Lastly, there is a low intake of eggs and sweets. The MD has been the interest of numerous studies, which have aimed to elaborate how exactly it affects the health of its consumers [11,12]. Higher adherence to this diet has shown many benefits to health, such as the reduced risk of developing diabetes, obesity, hypertension, cognitive diseases, hyperlipidemia and depression [13–18]. A multicenter trial in Spain showed that an MD with a higher intake of olive oil directly reduces risk of cardiovascular incidents, which is also in line with other studies that have shown the protective role of olive oil and the MD in several serious conditions, such as coronary disease and stroke [19–23].

A recent study by Iaccarino Idelson et al. showed that there is an association between physical activity and adherence to the MD, while they also have an inverse correlation with sedentary behavior [24]. Moreover, a systematic review with a meta-analysis by Malakou et al. found that the promotion of combined MD and physical activity showed a significant metabolic risk reduction [25]. In addition, the results of several Spanish studies have implied that the level of physical activity has a significant correlation with adherence to the MD [26–28]. However, MD could have some shortcomings in the physically active population, as the study by Passariello et al. has raised the question of whether MD can meet sufficient protein requirements [29]. Nevertheless, the results regarding the adherence to the MD and its connection to the level of physical activity in the population of regular fitness center users are still inconclusive.

Hence, the aim of this study was to investigate the association between physical activity and the adherence to the MD in fitness center users. Moreover, we aimed to evaluate the adherence to the specific dietary components of the MD and to assess the usage of dietary supplements and its association with MD in this population.

2. Materials and Methods

2.1. Study Design and Ethical Considerations

This cross-sectional survey-based study was performed among fitness center users in Split, Croatia, during the time period from July to October 2021. The restrictions in Croatia due to COVID-19 pandemic were lifted on February 15, and fitness centers have been operated normally since then.

The study was approved by the Ethics Committee of University of Split School of Medicine (No: 003-08/20-03/0005) and was conducted in regulation with the latest Helsinki declaration. The subjects gave consent to participate by submitting a completed questionnaire.

2.2. Participants

The study was conducted on fitness center users in Split, Croatia, using an online survey constructed with the Google Forms® application. The survey was distributed among fitness centers users using QR codes and emails, and through closed social media groups and fitness trainers.

Participation in the study was voluntary, and anonymity of the provided answers was guaranteed. The inclusion criteria were: 18–65 years of age, and using the fitness center for more than 3 months on at least a 1-time-per-week basis. The only exclusion criterion was involvement in professional sports. Professional sports were defined as involvement is all sport activities for which the participant receives payment and/or is competing in the professional tournaments, with the additional requirement that the subject is involved in that sport activity for ≥6 days/week.

2.3. *Questionnaires*

The survey consisted of three parts. The first part was the questionnaire which included general information about the participants such as the gender, age, anthropometric traits, frequency and duration of the training in the fitness center, involvement in professional sport and habits about usage of dietary supplements. The questionnaire included 12 items, and it was developed for the purpose of this study after extensive review of the available literature.

The second part of the survey was the Mediterranean Diet Serving Score (MDSS), a reliable, validated 14-item questionnaire with a verified Croatian version [30,31]. The MDSS is used to assess the adherence to the MD, and it is updated by the latest guidelines of the Mediterranean Diet Pyramid based on the frequency of consuming certain food and food groups and scoring them by one (1), two (2) or three (3) points depending on the recommendation on intake. Fruits, vegetables, cereals and olive oil are scored by three points, meaning they should be consumed every meal. The two-point foods are dairy products and nuts, which are recommended to be consumed daily. Lastly, one point goes to white meat, red meat, fish, potatoes, legumes, eggs, sweets and wine, which should be consumed once a week. The cutoff value for determining adherence to MD is a total MDSS score of ≥14 points.

The third part of the survey was the International Physical Activity Questionnaire Short Form (IPAQ-SF), an open-ended, reliable, validated questionnaire, which was verified in a Croatian version [32,33]. The IPAQ-SF evaluates the self-reported activity of four intensity levels: vigorous-intensity activities, moderate-intensity activities, walking and sitting [28–30]. It was suggested by the IPAQ-SF authors that, for observational studies, the "last 7 days recall" version should be used. MET (metabolic equivalent of task) minutes per week scores were calculated from the results of the IPAQ-SF according to the following formulas [34]:

- Walking MET-min/week = 3.3 × walking minutes × walking days
- Moderate MET-min/week = 4.0 × moderate activity minutes × moderate days
- Vigorous MET-min/week = 8.0 × vigorous activity minutes × vigorous days
- Total MET-min/week = walking + moderate + vigorous MET-min/week scores

2.4. *Survey Pre-Testing*

A survey pre-testing was conducted on a sample of 43 randomly chosen fitness center users. The average time needed to complete the survey was 12 min. The feedback from the responders showed that all the questions were clear and understandable. The final version of the survey consisted of 33 questions.

2.5. *Statistical Analyses*

All statistical analyses were performed using MedCalc for Microsoft Windows (MedCalc Software, Ostend, Belgium, version 17.4.1). Normality of distribution was evaluated using the Kolmogorov-Smirnov test. Continuous variables were presented as mean ± standard deviation or median (interquartile range) depending on the distribution normality. Categorical variables were presented as a whole number (N) with percentage (%). For determining differences between the groups, an independent samples t-test was used for continuous variables with normal distribution, whereas the Mann–Whitney U test was used for continuous variables with non-normal distribution. The chi-square (χ^2) test was used to determine differences between groups in terms of categorical variables. To investigate the correlation between variables, we used Spearman's rank correlation coefficient. Comparison of parameters between IPAQ-SF tertiles was performed using either one-way analysis of variance (ANOVA) with the post-hoc Tukey test or one-way analysis of variance on ranks with the post-hoc Dunn's test. A multiple linear regression analysis with a forward algorithm was applied to determine significant and independent correlates of the total MDSS score, which was defined as a dependent continuous variable. From these analyses, we reported the respective *p*-values with unstandardized β-coefficients, standard error and t-values. In addition, the independent predictors for adherence to the MD were evaluated with multivariable logistic regression, with the OR (odds ratio), 95% CI (95% confidence interval) and *p*-value reported. The level of statistical significance was set at *p*-value < 0.05.

3. Results

3.1. Baseline Characteristics

The study included 1220 participants, and there were 690 (56.5%) male and 530 (43.5%) female fitness center users. Their mean age was 29.1 ± 8.8 years. Most of them (52.6%) had the education level of master's degree, while the least (0.6%) had only elementary school. Furthermore, most of them (65.8%) were using dietary supplements, out of which whey protein was the most used supplement (76.8%) (Table 1).

Table 1. Baseline characteristics of the study sample and differences regarding gender.

Parameter	Study Sample n = 1220	Male n = 690	Female n = 530	p *
Age (years)	29.1 ± 8.8	28.2 ± 7.8	30.3 ± 9.9	0.001
Weight (kg)	79.3 ± 15.6	87.2 ± 13.0	68.9 ± 12.1	0.001
Height (cm)	179.6 ± 9.5	184.8 ± 7.3	172.0 ± 7.4	0.001
BMI (kg/m^2)	24.4 ± 3.4	25.4 ± 2.9	23.0 ± 3.6	0.001
Education level				
Elementary school (n, %)	7 (0.6)	2 (0.3%)	5 (0.9)	
High school (n, %)	315 (25.8)	172 (24.9)	143 (27.0)	0.334
Bachelor's degree (n, %)	256 (21.0)	143 (20.7)	113 (21.3)	
Master's degree (n, %)	642 (52.6)	373 (54.1)	269 (50.8)	
Using dietary supplements (n, %)				
Yes (n, %)	803 (65.8)	493 (71.4)	310 (58.5)	<0.001
No (n, %)	417 (34.2)	197 (28.6)	220 (41.5)	
Dietary supplements used				
Whey protein (n, %)	617 (76.8)	375 (76.0)	242 (78.0)	0.003
BCAA (n, %)	399 (49.6)	232 (47.0)	167 (53.8)	0.472
Creatine (n, %)	243 (30.2)	196 (39.7)	47 (15.1)	<0.001
Magnesium (n, %)	472 (58.7)	275 (55.7)	197 (63.5)	0.370
Vitamin C (n, %)	347 (43.2)	205 (41.5)	142 (45.8)	0.291
Vitamin B complex (n, %)	170 (21.1)	90 (18.2)	80 (25.8)	0.715
Multivitamin (n, %)	258 (32.1)	139 (28.1)	119 (38.3)	0.364
Duration of using a fitness center				
<1 year (n, %)	337 (27.6)	205 (29.7)	132 (24.9)	
1–3 years (n, %)	352 (28.9)	187 (27.1)	165 (31.1)	0.222
4–7 years (n, %)	232 (19.0)	128 (18.6)	104 (19.3)	
>7 years (n, %)	299 (24.5)	170 (24.6)	129 (24.3)	

All data are presented as whole numbers (percentage) or mean ± SD. Abbreviations: BMI—body mass index; MET—metabolic equivalent of task; BCAA—branched-chain amino acid. * Chi-square test or student *t*-test.

In regard to gender differences, male participants had a significantly higher weight (87.2 ± 13.0 vs. 68.9 ± 12.1 kg, $p < 0.001$), height (184.8 ± 7.3 vs. 172.0 ± 7.4 cm, $p < 0.001$) and BMI (25.4 ± 2.9 vs. 23.0 ± 3.6 kg/m^2, $p < 0.001$), while female participants were significantly older (28.2 ± 7.8 vs. 30.3 ± 9.9 years, $p < 0.001$). Furthermore, a significantly higher number of male participants were using dietary supplements (71.4% vs. 58.5%, $p < 0.001$) (Table 1).

We divided the study sample in tertiles using the IPAQ-SF results. The first tertile consisted of 407 participants, and their total MET min/week was <1750. The second tertile had 406 participants, and total MET min/week was 1750–3150, while the third tertile had 407 participants, and total MET min/week was >3150. There was a significant difference regarding gender in the three groups, as the first tertile had the lowest number of males (47.4%). Moreover, the first tertile had the highest number of high school education-level participant (32.7%), while the third tertile had the highest number of master's degree-level participants (58.0%). In addition, the third tertile had the highest usage of dietary supplements (73.7%). Out of the three groups, the third tertile also had the highest usage of whey (98.6%), BCAA (57.3%) and creatine (39.6%). Furthermore, the third tertile had the highest portion of participants who used the fitness center for more than 7 years (30.5%) (Table 2).

Table 2. Differences of the baseline characteristics between the tertiles of the IPAQ-SF results.

Parameter	First Tertile Group MET < 1750 min/Week n = 407	Second Tertile Group MET 1750–3150 min/Week n = 406	Third Tertile Group MET > 3150 min/Week n = 407	p *
Age (years)	29.0 ± 8.4	29.2 ± 9.3	29.1 ± 8.8	0.954
Male gender (n, %)	193 (47.4%)	255 (62.8%)	242 (59.5%)	<0.001
Weight (kg)	79.9 ± 15.8	77.7 ± 14.4	79.9 ± 16.2	0.095
Height (cm)	179.9 ± 9.7	179.0 ± 9.3	179.7 ± 9.5	0.473
BMI (kg/m^2)	24.4 ± 3.3	24.0 ± 3.1	24.5 ± 3.8	0.114
Education level				
Elementary school (n, %)	2 (0.5)	2 (0.5)	3 (0.7)	<0.001
High school (n, %)	133 (32.7)	88 (21.7)	94 (23.1)	
Bachelor's degree (n, %)	69 (17.0)	113 27.8)	74 (18.2)	
Master's degree (n, %)	203 (49.9)	203 (50.0)	236 (58.0)	
Using dietary supplements (n, %)				
Yes (n, %)	229 (56.3)	274 (67.5)	300 (73.7)	<0.001
No (n, %)	178 (43.7)	132 (32.5)	107 (26.3)	
Dietary supplements used				
Whey protein (n, %)	115 (50.2)	206 (89.9)	296 (98.6)	<0.001
BCAA (n, %)	73 (31.8)	154 (67.2)	172 (57.3)	<0.001
Creatine (n, %)	22 (9.6)	102 (44.5)	119 (39.6)	<0.001
Magnesium (n, %)	144 (62.8)	165 (72.0)	163 (71.1)	0.240
Vitamin C (n, %)	118 (51.5)	112 (48.9)	117 (51.0)	0.893
Vitamin B complex (n, %)	59 (25.7)	56 (24.4)	55 (24.0)	0.916
Multivitamin (n, %)	84 (36.6)	90 (39.3)	84 (36.6)	0.827
Duration of using a fitness center				
<1 year (n, %)	98 (24.1)	118 (29.1)	121 (29.7)	<0.001
1–3 years (n, %)	137 (33.7)	115 (28.3)	100 (24.6)	
4–7 years (n, %)	76 (18.7)	94 (23.2)	62 (15.2)	
>7 years (n, %)	96 (23.6)	79 (19.5)	124 (30.5)	

All data are presented as whole numbers (percentage) or mean ± SD. Abbreviations: BMI—body mass index; MET—metabolic equivalent of task; BCAA—branched-chain amino acid. * Chi-square test or one-way analysis of variance.

3.2. MDSS Results in the Study Sample

The MDSS score in the whole study sample was 8.0 (5.0–12.0), and a total of 227 (18.6%) participants were adherent to the MD (total MDSS score ≥ 14) (Figure 1). Regarding the components of the MDSS, the highest adherence was in the consumption of potatoes (84.3%) and white meat (82.4%), while the lowest adherence was in the wine consumption (8.0%) (Table 3). Moreover, there was a significant positive correlation between the total MDSS score and the total MET min/week ($r = 0.302$, $p < 0.001$) (Figure 2).

In regard to the gender differences, female participants had a significantly higher adherence in the consumption of sweets (305 (57.5%) vs. 357 (51.7%), $p = 0.049$) and fruits (161 (30.4%) vs. 165 (23.9%), $p = 0.013$). There were no significant differences regarding the adherence to the other MDSS components (Table S1). In addition, the female participants had a statistically higher total MDSS score compared to the male participants (8.5(6.0–13.0) vs. 8.0(5.0–12.0), $p = 0.041$) (Figure 3). However, when comparing the adherence to the MD (total MDSS score ≥ 14), there was no statistically significant difference between the genders ($p = 0.811$) (Figure 4).

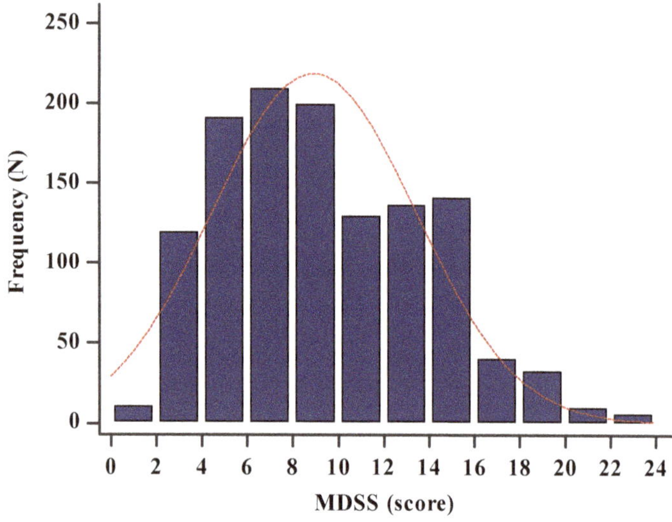

Figure 1. Histogram showing the MDSS score in the study sample (n = 1220).

Table 3. Differences in the adherence to the MDSS components between the tertiles of the IPAQ-SF results.

Parameter	First Tertile Group MET < 1750 min/Week n = 407	Second Tertile Group MET 1750–3150 min/Week n = 406	Third Tertile Group MET > 3150 min/Week n = 407	p *
Cereals (n, %)	87 (21.4)	93 (23.0)	156 (38.3)	<0.001
Potatoes (n, %)	367 (90.2)	343 (84.7)	318 (78.1)	<0.001
Olive oil (n, %)	59 (14.5)	85 (21.0)	119 (29.2)	<0.001
Nuts (n, %)	124 (30.5)	174 (43.0)	171 (42.0)	<0.001
Fruits (n, %)	70 (17.2)	98 (24.1)	158 (38.8)	<0.001
Vegetables (n, %)	91 (22.4)	111 (27.3)	178 (43.7)	<0.001
Dairy (n, %)	64 (15.7)	118 (29.1)	154 (37.8)	<0.001
Legumes (n, %)	313 (76.9)	248 (61.2)	270 (66.3)	<0.001
Eggs (n, %)	224 (55.0)	186 (46.0)	195 (47.9)	0.025
Fish (n, %)	242 (59.5)	217 (53.7)	266 (65.4)	0.003
White meat (n, %)	372 (91.4)	304 (74.9)	329 (80.8)	<0.001
Red meat (n, %)	73 (17.9)	145 (35.8)	187 (45.9)	<0.001
Sweets (n, %)	206 (50.6)	265 (65.3)	191 (46.9)	<0.001
Wine (n, %)	1 (0.2)	59 (14.6)	38 (9.3)	<0.001

All data are presented as whole numbers (percentage). * Chi-square test.

There were statistically significant differences in the adherence to MDSS components between the tertiles of the IPAQ-SF results (Table 3). In addition, there was a significant difference between the tertiles regarding the total MDSS score (H = 82.391, $p < 0.001$) (Figure 5). The post-hoc Dunn's test analysis showed that there was a significant difference between all three tertiles (first tertile: 7.0 (5.0–9.0), second tertile: 8.0 (5.0–12.0), third tertile: 11.0 (6.0–14.0); $p < 0.05$) (Figure 4). Furthermore, after comparison of the adherence to the MD (total MDSS score $\geq$ 14), there was a significant difference between the tertiles ($p < 0.001$). Most participants who were adherent to the MD were in the third tertile (34.4%), while the least were in the first tertile (6.1%) (Figure 4).

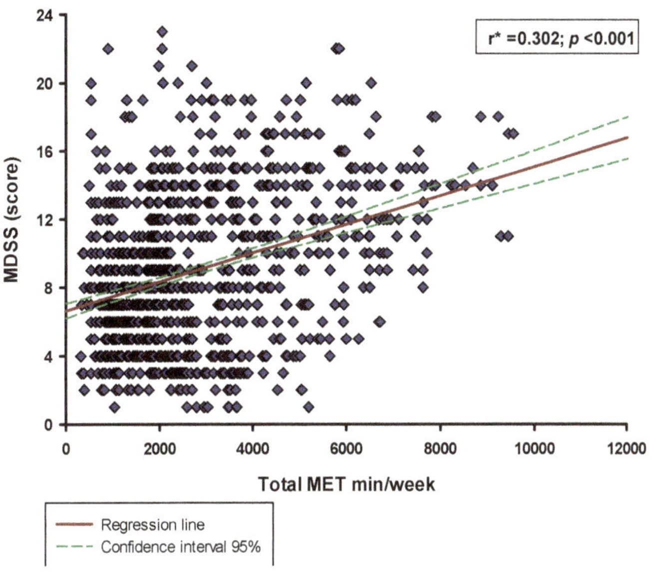

Figure 2. Correlation between the total MDSS score and the total MET min/week in the whole study sample (n = 1220). * Spearman's correlation coefficient.

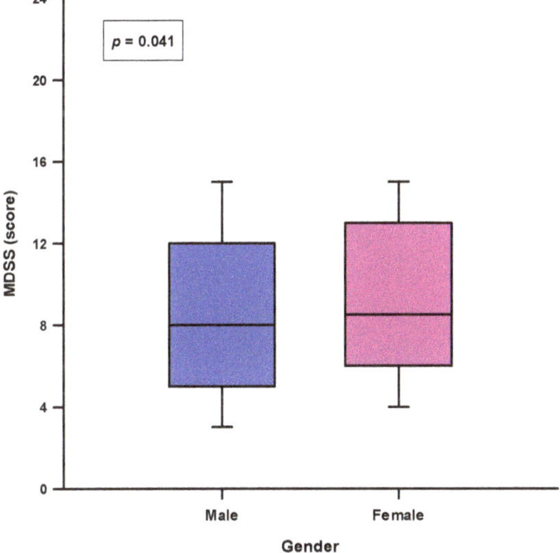

Figure 3. Difference of the total MDDS score between the males (n = 690) and females (n = 530). Tested with the Mann-Whitney U test.

In addition, after comparison of the durations of using a fitness center, there was a statistically significant difference between the groups (H = 13.685, p = 0.003) in the total MDSS score. The post-hoc Dunn's test showed that the participants who used the fitness center for less than 1 year had the lowest total MDSS score and were significantly different (p <0.05) from the other three groups (<1 year: 7.0 (4.0–12.0); 1–3 years: 9.0 (6.0–12.75); 4–7 years: 8.0 (6.0–13.0); >7 years: 9.0 (6.0–12.0)).

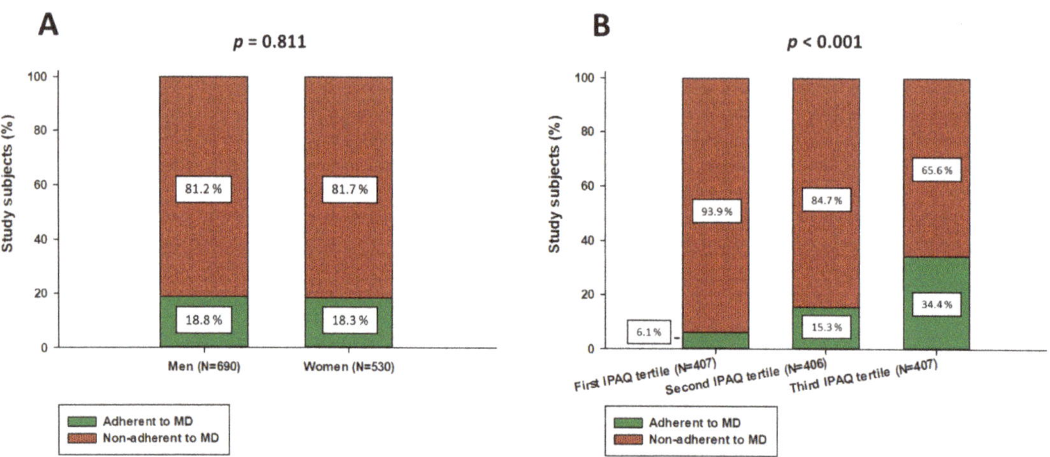

Figure 4. Differences in the adherence to the MD between (**A**) gender and (**B**) tertiles of the IPAQ-SF results. The chi-square test was used for analysis.

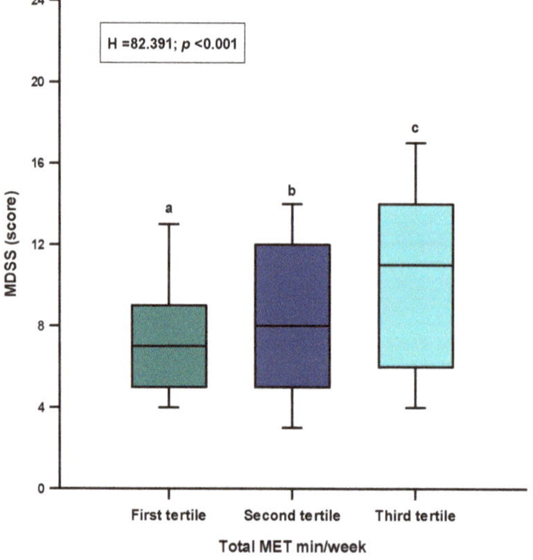

Figure 5. Difference of the total MDSS score between the tertiles of the IPAQ-SF results. Tested using one-way analysis of variance on ranks with the post-hoc Dunn's test to examine the differences between each group. a vs. b = $p < 0.05$; a vs. c = $p < 0.05$; a vs. d = $p < 0.05$; d vs. b = $p < 0.05$; d vs. c = $p < 0.05$.

3.3. Regression Analyses

Multiple linear regression analysis showed that the total MDSS score retained significant association with the total MET min/week ($\beta \pm$ SE, 0.007 $\pm$ 0.0006, $p < 0.001$) and the duration of using a fitness center (-0.401 ± 0.102, $p = 0.001$) after the model adjustment for age and BMI, with the total MDSS score as a dependent variable (Table 4).

Table 4. Multiple linear regression model of the independent predictors of the total MDSS score.

Variable.	β *	SE	t-Value	*p*
Age	−0.003	0.013	−0.222	0.824
BMI	0.017	0.038	0.464	0.642
Total MET min/week	0.007	0.0006	11.509	<0.001
Duration of using a fitness center	0.401	0.102	3.912	0.001

Abbreviations: SE—standard error; BMI—body mass index; MET—metabolic equivalent of task. * unstandardized coefficient β.

Furthermore, multivariable logistic regression showed that the third and fourth quartile of MET min/week ($p < 0.001$), dietary supplements usage ($p = 0.023$) and the duration of using a fitness center for 1–3 years ($p = 0.021$) were significant predictors of positive adherence to the MD when computed along with the baseline characteristics (Table 5).

Table 5. Multivariable logistic regression analysis of the independent predictors for positive adherence to the MD according to the total MDSS score.

Variable	aOR [95% CI]	*p*
Female sex [1]	1.09 [0.77, 1.54]	0.811
Using the fitness center for 1–3 years [2]	1.93 [1.24, 3.00]	0.021
Using the fitness center for 4–7 years [2]	1.47 [0.97, 2.23]	0.071
Using the fitness center for >7 years [2]	1.05 [0.68, 1.63]	0.814
Dietary supplements usage [3]	1.52 [1.06, 2.17]	0.023
Older age	1.00 [0.98, 1.02]	0.874
Total MET min/week 2nd quartile [4]	1.61 [0.88, 2.96]	0.126
Total MET min/week 3rd quartile [4]	3.94 [2.28, 6.80]	<0.001
Total MET min/week 4th quartile [4]	8.08 [4.78, 13.67]	<0.001
BMI	0.99 [0.95, 1.05]	0.936

[1] Reference group are male subjects. [2] Reference group are subjects with the shortest gym attendance (<1 year). [3] Reference group are subjects not utilizing dietary supplements. [4] Reference group are subjects within the 1st MET quartile. Abbreviations: MD—Mediterranean diet; OR—multivariable adjusted odds ratio; 95% CI—95% confidence interval; BMI—body mass index; MET—metabolic equivalent of task.

4. Discussion

The findings of this study showed that 18.6% of fitness center users were adherent to the MD, and there was a significant positive correlation between the level of physical activity and the MDSS score in this population. Moreover, after dividing the sample into tertiles based on the IPAQ-SF score, the third tertile (MET > 3150 min/wk) had the most fitness center users adherent to the MD, while the first tertile (MET < 1750 min/wk) had the least. With thorough research of the available literature, there are no other studies which have investigated the adherence to MD in this specific population.

Previous studies have shown that the level of fitness is significantly associated with the adherence to the MD [24,26–28]. A recent Spanish study on university students determined that there is a significant association between adherence to the MD and both a high level of muscular fitness and high level of cardiorespiratory fitness [28]. Furthermore, two studies conducted on schoolchildren found that subjects who were more physically active also had a higher adherence to the MD [26,27]. The biggest difference of the present study from the aforementioned ones is the studied population. Specifically, our studied population exclusively consisted of adults, which enabled more reliable inferences, as the results acquired in the pediatric population may be confounded by the effect of parental care on both dietary preferences and physical activity. Nevertheless, these results emphasize the possible connection between physical activity and the adherence to the MD. Even though the benefits of the MD are well established, studies show that the prevalence of the adherent subjects among the general population are very low, even in Mediterranean countries [35–37]. A study conducted on Spanish and Romanian students showed higher adherence to the MD in students who distinguished the importance of proper nutrition to achieve better health and sport results [38]. Moreover, a longitudinal study conducted on healthy adolescents showed that that the adherence to the MD significantly increased after nutrition education sessions [39]. These outcomes imply that the low adherence to the MD could be due to the lack of awareness about the benefits of this diet even if it is the traditional cuisine of the participants' region. However, since this type of diet also minimizes meat and meat products, which are the most protein-rich type of nutrition, it is very peculiar why individuals with a high physical activity are more adherent to the MD. Our results showed that individuals with higher physical activity also use more dietary supplements, so it is possible that they meet their protein requirements through supplementation. A recent Italian study similarly showed that Mediterranean athletes used dietary supplements, but they also determined that the individuals who used supplements were already making food consumption choices that would guarantee them an adequate amount of protein intake [29].

After dividing the population into tertiles based on the IPAQ-SF results, there was a significant difference between these groups regarding the total MDSS score and all the distinctive dietary components of the MD. The third tertile, which had the highest physical activity, also had the highest MDSS scores, as well as the highest number of subjects adherent to the MD. In regard to specific dietary components, the third tertile also had the highest adherence to cereals, olive oil, fruits, vegetables, dairy, fish and red meat. In contrast, the first tertile, which had the lowest physical activity, also had the lowest MDSS scores, as well as the lowest number of subjects adherent to the MD. However, in regard to specific dietary components of the MD, they had the highest adherence to the potatoes, legumes, eggs and white meat. These results are interesting since the MD, as aforementioned, de-emphasizes the consumption of meat, which is usually the main source of proteins for physically active individuals. It has been well established that both professional athletes and physically active non-athletes are more prone to use white meat as a source of proteins than red meat. Knowing that better adherence to MD in terms of meat, in fact, means less intake of it, it is not surprising that participants with higher degree of physical activity were less adherent to white meat intake. This subgroup simply ingested more white meat and were thus less adherent. On the other hand, participants with higher degree of physical activity were more adherent to red meat because they ingested less red meat on average. In addition, these results could also be interpreted regarding the goals of fitness center users. Nowadays, most fitness center users are, in addition to health improvement, aiming to achieve a desired level of body composition [40]. This is especially present among adolescents and young adults, which comprised a major percentage of our study population. Furthermore, younger physically active individuals are usually also more burdened with following a healthy diet they deem fit for reaching their aims. MD would possibly be their best choice since it is currently considered one of the healthiest dietary models worldwide. However, these findings need to be addressed in future studies.

Even though there was no significant difference between genders in the number of participants adherent to the MD, the females had a significantly higher total MDSS score. This result is in line with other studies, which have shown that females are more adherent to the MD [41,42]. A study conducted on Portuguese adults showed that women have a significantly higher adherence to the MD, as well as a significantly higher number of meals per day [43]. Moreover, a recent study conducted on Italian adults showed that being female, as well as having a higher income and education level, were the most relevant factors influencing the probability of having a higher adherence to the MD [44]. Some of the possible reasons are that women tend to consume more fruit and vegetables, less meat and, in general, easily implement healthier eating patterns [45]. Another possible explanation could be that females have a tendency of being more aware about the type and quality of their diet. However, due to a specific population of our study and the different geographical distribution of the included participants, it is difficult to compare our results with these other studies.

Out of the spotlight but also an important finding of this study is the difference regarding the total MDSS score between the different durations of using a fitness center.

Our results show that participants who used the fitness center for less than 1 year also had the lowest total MDSS score compared to those who had used it for 1–3 years, 4–7 years and over 7 years. These results imply that the "newcomers" to the fitness center are significantly less adherent to the MD. However, since those who used the fitness center for longer than 1 year had significantly higher MDSS scores, we could hypothesize that, over time, the newcomers become more adherent to the MD. This could be due to their adaptation to the fitness lifestyle, which includes a more balanced and healthy diet. As aforementioned, regular fitness could possibly be associated with a healthier diet [46]. However, this finding needs to be addressed more thoroughly, especially regarding other factors which possibly influence this gradual adaptation to the MD in new fitness center users.

There are several limitations to this study. Its cross-sectional design restricts the possibility of causal conclusions. Moreover, the study was conducted in only one city in Croatia, so it is possible that these results are region-specific. Since the main tool to assess the evaluated parameters was a questionnaire, there is a possibility that the subjects had a recall bias or had an excess of subjectivity in some of the answers. Furthermore, due to the self-administration of the questionnaire, it is possible that the participants had biased or unreliable answers. Lastly, our sample mostly included a younger population, which may have interfered with the results.

5. Conclusions

In conclusion, this study showed that almost every fifth fitness center user in our sample was adherent to the MD. Moreover, the level of physical activity showed a significant positive correlation with the adherence to the MD. However, gender did not seem to be a strong factor in the adherence to the MD in this population. Whereas females did have a higher total MDSS score, there still was not any significant difference in the number of adherent subjects between genders. Lastly, these results showed that physical activity is also associated with dietary supplements consumption, as 82.3% of subjects in the group with the highest MET min/wk used some sort of dietary supplementation. These outcomes emphasize the importance of physical activity as they imply that, with higher levels of physical activity, people are also possibly more aware of the importance that a healthy and balanced diet has on their well-being.

Supplementary Materials: The following are available online at https://www.mdpi.com/article/10.3390/nu13114038/s1, Table S1: Adherance to the MDSS components and differences regarding gender.

Author Contributions: Conceptualization, D.M. and J.B.; methodology, D.M., J.B. and D.T.; software, D.T., M.K., D.R., J.V. and T.T.K.; validation, D.M., D.T., M.V., S.L.-F. and J.B.; formal analysis, D.M., D.R., M.V. and J.V.; investigation, D.M., L.M., D.R., I.M. and S.L.-F.; resources, J.B.; data curation, D.M., D.T., L.M., M.V. and I.M.; writing—original draft preparation, D.M., D.T., L.M., M.V. and D.R.; writing—review and editing, D.M., D.T., M.K., I.M. and T.T.K.; visualization, L.M., M.K., J.V., I.M. and S.L.-F.; supervision, D.M., T.T.K. and J.B.; project administration, D.M. and J.B.; funding acquisition, J.B. All authors have read and agreed to the published version of the manuscript.

Funding: This research received no external funding.

Institutional Review Board Statement: The study was conducted according to the guidelines of the Declaration of Helsinki and approved by the Ethics Committee of the University of Split School of Medicine (Number: 003-08/20-03/0005; 3 July 2020).

Informed Consent Statement: Informed consent was obtained from all subjects involved in the study.

Data Availability Statement: All the data is available upon request to the corresponding author.

Acknowledgments: In this section, you can acknowledge any support given which is not covered by the author contribution or funding sections. This may include administrative and technical support, or donations in kind (e.g., materials used for experiments).

Conflicts of Interest: The authors declare no conflict of interest.

References

1. Ruegsegger, G.N.; Booth, F.W. Health Benefits of Exercise. *Cold Spring Harb. Perspect. Med.* **2018**, *8*, a029694. [CrossRef]
2. Block, L.G.; Grier, S.A.; Childers, T.L.; Davis, B.; Ebert, J.E.J.; Kumanyika, S.; Laczniak, R.N.; Machin, J.E.; Motley, C.M.; Peracchio, L.; et al. From Nutrients to Nurturance: A Conceptual Introduction to Food Well-Being. *J. Public Policy Mark.* **2011**, *30*, 5–13. [CrossRef]
3. Renner, B.; Sproesser, G.; Strohbach, S.; Schupp, H.T. Why we eat what we eat. The Eating Motivation Survey (TEMS). *Appetite* **2012**, *59*, 117–128. [CrossRef] [PubMed]
4. Yang, Y.J. An Overview of Current Physical Activity Recommendations in Primary Care. *Korean J. Fam. Med.* **2019**, *40*, 135–142. [CrossRef] [PubMed]
5. Booth, F.W.; Roberts, C.K.; Laye, M.J. Lack of exercise is a major cause of chronic diseases. *Compr. Physiol.* **2012**, *2*, 1143–1211.
6. Sharma, A.; Madaan, V.; Petty, F.D. Exercise for mental health. *Prim. Care Companion J. Clin. Psychiatry* **2006**, *8*, 106. [CrossRef] [PubMed]
7. Martinovic, D.; Tokic, D.; Vilovic, M.; Rusic, D.; Bukic, J.; Bozic, J. Sport Dietary Supplements and Physical Activity in Biomedical Students. *Int. J. Environ. Res. Public Health* **2021**, *18*, 2046. [CrossRef]
8. Gardiner, P.; Kemper, K.J.; Legedza, A.; Phillips, R.S. Factors associated with herb and dietary supplement use by young adults in the United States. *BMC Complement. Altern. Med.* **2007**, *7*, 39. [CrossRef]
9. Cena, H.; Calder, P.C. Defining a Healthy Diet: Evidence for the Role of Contemporary Dietary Patterns in Health and Disease. *Nutrients* **2020**, *12*, 334. [CrossRef]
10. Davis, C.; Bryan, J.; Hodgson, J.; Murphy, K. Definition of the Mediterranean Diet; A Literature Review. *Nutrients* **2015**, *7*, 9139–9153. [CrossRef]
11. Augimeri, G.; Bonofiglio, D. The Mediterranean Diet as a Source of Natural Compounds: Does It Represent a Protective Choice against Cancer? *Pharmaceuticals* **2021**, *14*, 920. [CrossRef] [PubMed]
12. Augimeri, G.; Montalto, F.I.; Giordano, C.; Barone, I.; Lanzino, M.; Catalano, S.; Andò, S.; De Amicis, F.; Bonofiglio, D. Nutraceuticals in the Mediterranean Diet: Potential Avenues for Breast Cancer Treatment. *Nutrients* **2021**, *13*, 2557. [CrossRef]
13. Zappalà, G.; Buscemi, S.; Mulè, S.; La Verde, M.; D'Urso, M.; Corleo, D.; Marranzano, M. High adherence to Mediterranean diet, but not individual foods or nutrients, is associated with lower likelihood of being obese in a Mediterranean cohort. *Eat. Weight Disord.* **2018**, *23*, 605–614. [CrossRef] [PubMed]
14. La Verde, M.; Mulè, S.; Zappalà, G.; Privitera, G.; Maugeri, G.; Pecora, F.; Marranzano, M. Higher adherence to the Mediterranean diet is inversely associated with having hypertension: Is low salt intake a mediating factor? *Int. J. Food Sci. Nutr.* **2018**, *69*, 235–244. [CrossRef] [PubMed]
15. Psaltopoulou, T.; Sergentanis, T.N.; Panagiotakos, D.B.; Sergentanis, I.N.; Kosti, R.; Scarmeas, N. Mediterranean diet, stroke, cognitive impairment, and depression: A meta-analysis. *Ann. Neurol.* **2013**, *74*, 580–591. [CrossRef]
16. Loughrey, D.G.; Lavecchia, S.; Brennan, S.; Lawlor, B.A.; Kelly, M.E. The Impact of the Mediterranean Diet on the Cognitive Functioning of Healthy Older Adults: A Systematic Review and Meta-Analysis. *Adv. Nutr.* **2017**, *8*, 571–586. [CrossRef] [PubMed]
17. Platania, A.; Zappala, G.; Mirabella, M.U.; Gullo, C.; Mellini, G.; Beneventano, G.; Maugeri, G.; Marranzano, M. Association between Mediterranean diet adherence and dyslipidaemia in a cohort of adults living in the Mediterranean area. *Int. J. Food Sci. Nutr.* **2018**, *69*, 608–618. [CrossRef]
18. Guasch-Ferré, M.; Merino, J.; Sun, Q.; Fitó, M.; Salas-Salvadó, J. Dietary Polyphenols, Mediterranean Diet, Prediabetes, and Type 2 Diabetes: A Narrative Review of the Evidence. *Oxid. Med. Cell. Longev.* **2017**, *2017*, 6723931. [CrossRef]
19. Bendinelli, B.; Masala, G.; Saieva, C.; Salvini, S.; Calonico, C.; Sacerdote, C.; Agnoli, C.; Grioni, S.; Frasca, G.; Mattiello, A.; et al. Fruit, vegetables, and olive oil and risk of coronary heart disease in Italian women: The EPICOR Study. *Am. J. Clin. Nutr.* **2011**, *93*, 275–283. [CrossRef]
20. Estruch, R.; Martínez-González, M.A.; Corella, D.; Salas-Salvadó, J.; Ruiz-Gutiérrez, V.; Covas, M.I.; Fiol, M.; Gómez-Gracia, E.; López-Sabater, M.C.; Vinyoles, E.; et al. Effects of a Mediterranean-style diet on cardiovascular risk factors: A randomized trial. *Ann. Intern. Med.* **2006**, *145*, 1–11. [CrossRef]
21. Buckland, G.; Travier, N.; Barricarte, A.; Ardanaz, E.; Moreno-Iribas, C.; Sánchez, M.J.; Molina-Montes, E.; Chirlaque, M.D.; Huerta, J.M.; Navarro, C.; et al. Olive oil intake and CHD in the European Prospective Investigation into Cancer and Nutrition Spanish cohort. *Br. J. Nutr.* **2012**, *108*, 2075–2082. [CrossRef]
22. Samieri, C.; Féart, C.; Proust-Lima, C.; Peuchant, E.; Tzourio, C.; Stapf, C.; Berr, C.; Barberger-Gateau, P. Olive oil consumption, plasma oleic acid, and stroke incidence: The Three-City Study. *Neurology* **2011**, *77*, 418–425. [CrossRef]
23. Estruch, R.; Ros, E.; Salas-Salvadó, J.; Covas, M.I.; Corella, D.; Arós, F.; Gómez-Gracia, E.; Ruiz-Gutiérrez, V.; Fiol, M.; Lapetra, J.; et al. Primary Prevention of Cardiovascular Disease with a Mediterranean Diet Supplemented with Extra-Virgin Olive Oil or Nuts. *N. Engl. J. Med.* **2018**, *378*, 13. [CrossRef]
24. Iaccarino Idelson, P.; Scalfi, L.; Valerio, G. Adherence to the Mediterranean Diet in children and adolescents: A systematic review. *Nutr. Metab. Cardiovasc. Dis.* **2017**, *27*, 283–299. [CrossRef]
25. Malakou, E.; Linardakis, M.; Armstrong, M.E.G.; Zannidi, D.; Foster, C.; Johnson, L.; Papadaki, A. The Combined Effect of Promoting the Mediterranean Diet and Physical Activity on Metabolic Risk Factors in Adults: A Systematic Review and Meta-Analysis of Randomised Controlled Trials. *Nutrients* **2018**, *10*, 1577. [CrossRef]

26. Fernández-Iglesias, R.; Álvarez-Pereira, S.; Tardón, A.; Fernández-García, B.; Iglesias-Gutiérrez, E. Adherence to the Mediterranean Diet in a School Population in the Principality of Asturias (Spain): Relationship with Physical Activity and Body Weight. *Nutrients* **2021**, *13*, 1507. [CrossRef] [PubMed]
27. López-Gil, J.F.; Brazo-Sayavera, J.; García-Hermoso, A.; Yuste Lucas, J.L. Adherence to Mediterranean Diet Related with Physical Fitness and Physical Activity in Schoolchildren Aged 6–13. *Nutrients* **2020**, *12*, 567. [CrossRef]
28. Galan-Lopez, P.; Sánchez-Oliver, A.J.; Ries, F.; González-Jurado, J.A. Mediterranean Diet, Physical Fitness and Body Composition in Sevillian Adolescents: A Healthy Lifestyle. *Nutrients* **2019**, *11*, 2009. [CrossRef] [PubMed]
29. Passariello, C.L.; Marchionni, S.; Carcuro, M.; Casali, G.; Della Pasqua, A.; Hrelia, S.; Malaguti, M.; Lorenzini, A. The Mediterranean Athlete's Nutrition: Are Protein Supplements Necessary? *Nutrients* **2020**, *12*, 3681. [CrossRef] [PubMed]
30. Monteagudo, C.; Mariscal-Arcas, M.; Rivas, A.; Lorenzo-Tovar, M.L.; Tur, J.A.; Olea-Serrano, F. Proposal of a Mediterranean Diet Serving Score. *PLoS ONE* **2015**, *10*, e0128594. [CrossRef] [PubMed]
31. Marendić, M.; Polić, N.; Matek, H.; Oršulić, L.; Polašek, O.; Kolčić, I. Mediterranean diet assessment challenges: Validation of the Croatian Version of the 14-item Mediterranean Diet Serving Score (MDSS) Questionnaire. *PLoS ONE* **2021**, *16*, e0247269. [CrossRef]
32. Sember, V.; Meh, K.; Sorić, M.; Starc, G.; Rocha, P.; Jurak, G. Validity and Reliability of International Physical Activity Questionnaires for Adults across EU Countries: Systematic Review and Meta Analysis. *Int. J. Environ. Res. Public Health* **2020**, *17*, 7161. [CrossRef] [PubMed]
33. Craig, C.L.; Marshall, A.L.; Sjöström, M.; Bauman, A.E.; Booth, M.L.; Ainsworth, B.E.; Pratt, M.; Ekelund, U.; Yngve, A.; Sallis, J.F.; et al. International physical activity questionnaire: 12-country reliability and validity. *Med. Sci. Sports Exerc.* **2003**, *35*, 1381–1395. [CrossRef]
34. Ainsworth, B.E.; Haskell, W.L.; Whitt, M.C.; Irwin, M.L.; Swartz, A.M.; Strath, S.J.; O'Brien, W.L.; Bassett, D.R., Jr.; Schmitz, K.H.; Emplaincourt, P.O.; et al. Compendium of physical activities: An update of activity codes and MET intensities. *Med. Sci. Sports Exerc.* **2000**, *32*, S498–S504. [CrossRef] [PubMed]
35. La Fauci, V.; Alessi, V.; Assefa, D.Z.; Lo Giudice, D.; Calimeri, S.; Ceccio, C.; Antonuccio, G.M.; Genovese, C.; Squeri, R. Mediterranean diet: Knowledge and adherence in Italian young people. *Clin. Ter.* **2020**, *171*, e437–e443. [PubMed]
36. Dinu, M.; Lotti, S.; Pagliai, G.; Pisciotta, L.; Zavatarelli, M.; Borriello, M.; Solinas, R.; Galuffo, R.; Clavarino, A.; Acerra, E.; et al. Mediterranean Diet Adherence in a Sample of Italian Adolescents Attending Secondary School—The "#facciamoComunicAzione" Project. *Nutrients* **2021**, *13*, 2806. [CrossRef]
37. Kyriacou, A.; Evans, J.M.; Economides, N. Adherence to the Mediterranean diet by the Greek and Cypriot population: A systematic review. *Eur. J. Public Health* **2015**, *25*, 1012–1018. [CrossRef]
38. Chacón-Cuberos, R.; Badicu, G.; Zurita-Ortega, F.; Castro-Sánchez, M. Mediterranean Diet and Motivation in Sport: A Comparative Study Between University Students from Spain and Romania. *Nutrients* **2018**, *11*, 30. [CrossRef]
39. Morelli, C.; Avolio, E.; Galluccio, A.; Caparello, G.; Manes, E.; Ferraro, S.; Caruso, A.; De Rose, D.; Barone, I.; Adornetto, C.; et al. Nutrition Education Program and Physical Activity Improve the Adherence to the Mediterranean Diet: Impact on Inflammatory Biomarker Levels in Healthy Adolescents From the DIMENU Longitudinal Studby. *Front. Nutr.* **2021**, *8*, 685247. [CrossRef]
40. Ouyang, Y.; Wang, K.; Zhang, T.; Peng, L.; Song, G.; Luo, J. The Influence of Sports Participation on Body Image, Self-Efficacy, and Self-Esteem in College Students. *Front. Psychol.* **2020**, *10*, 3039. [CrossRef]
41. Bonaccorsi, G.; Furlan, F.; Scocuzza, M.; Lorini, C. Adherence to Mediterranean Diet among Students from Primary and Middle School in the Province of Taranto, 2016–2018. *Int. J. Environ. Res. Public Health* **2020**, *17*, 5437. [CrossRef] [PubMed]
42. Rosi, A.; Calestani, M.V.; Parrino, L.; Milioli, G.; Palla, L.; Volta, E.; Brighenti, F.; Scazzina, F. Weight Status Is Related with Gender and Sleep Duration but Not with Dietary Habits and Physical Activity in Primary School Italian Children. *Nutrients* **2017**, *9*, 579. [CrossRef]
43. Andrade, V.; Jorge, R.; García-Conesa, M.-T.; Philippou, E.; Massaro, M.; Chervenkov, M.; Ivanova, T.; Maksimova, V.; Smilkov, K.; Ackova, D.G.; et al. Mediterranean Diet Adherence and Subjective Well-Being in a Sample of Portuguese Adults. *Nutrients* **2020**, *12*, 3837. [CrossRef] [PubMed]
44. Biasini, B.; Rosi, A.; Menozzi, D.; Scazzina, F. Adherence to the Mediterranean Diet in Association with Self-Perception of Diet Sustainability, Anthropometric and Sociodemographic Factors: A Cross-Sectional Study in Italian Adults. *Nutrients* **2021**, *13*, 3282. [CrossRef] [PubMed]
45. Gregório, M.J.; Rodrigues, A.M.; Graça, P.; de Sousa, R.D.; Dias, S.S.; Branco, J.C.; Canhão, H. Food Insecurity Is Associated with Low Adherence to the Mediterranean Diet and Adverse Health Conditions in Portuguese Adults. *Front. Public Health* **2018**, *6*, 38. [CrossRef]
46. Joo, J.; Williamson, S.A.; Vazquez, A.I.; Fernandez, J.R.; Bray, M.S. The influence of 15-week exercise training on dietary patterns among young adults. *Int. J. Obes.* **2019**, *43*, 1681–1690. [CrossRef]

MDPI

Article

The Role of Socioeconomic Status in Adherence to the Mediterranean Diet and Body Mass Index Change: A Follow-Up Study in the General Population of Southern Croatia

Ajka Pribisalić [1], Romana Popović [2], Fiorella Pia Salvatore [3], Maja Vatavuk [4], Marija Mašanović [5], Caroline Hayward [6], Ozren Polašek [1,7] and Ivana Kolčić [1,7,*]

1 Department of Public Health, University of Split School of Medicine, Šoltanska 2, 21000 Split, Croatia; ajka.pribisalic@mefst.hr (A.P.); opolasek@mefst.hr (O.P.)
2 NUTRITIUS—Nutrition Counseling, Primorska 30, 20000 Dubrovnik, Croatia; nutritius.romana@gmail.com
3 Department of Economics, University of Foggia, 71121 Foggia, Italy; fiorellapia.salvatore@unifg.it
4 Department of Cardiac Surgery, University Hospital of Split, 21000 Split, Croatia; majavatavuk1@gmail.com
5 Department for Social Medicine, Division for Health Promotion, Public Health Institute of Dubrovnik Neretva County, Dr. A. Šercera 4a, pp 58, 20001 Dubrovnik, Croatia; marija.masanovic@zzjzdnz.hr
6 MRC Human Genetics Unit, Institute of Genetics and Molecular Medicine, University of Edinburgh, Edinburgh EH8 9YL, UK; caroline.hayward@ed.ac.uk
7 Algebra LAB, Algebra University College, Ilica 242, 10000 Zagreb, Croatia
* Correspondence: ikolcic@mefst.hr; Tel.: +385-91-576-2263

Citation: Pribisalić, A.; Popović, R.; Salvatore, F.P.; Vatavuk, M.; Mašanović, M.; Hayward, C.; Polašek, O.; Kolčić, I. The Role of Socioeconomic Status in Adherence to the Mediterranean Diet and Body Mass Index Change: A Follow-Up Study in the General Population of Southern Croatia. *Nutrients* **2021**, *13*, 3802. https://doi.org/10.3390/nu13113802

Academic Editor: Daniela Bonofiglio

Received: 15 September 2021
Accepted: 25 October 2021
Published: 26 October 2021

Abstract: The Mediterranean diet (MD) is one of the most healthful dietary patterns, beneficial for humans and the environment. However, the MD has recently exhibited a declining trend, especially in younger and less affluent people. This study investigated the association between socioeconomic indicators and adherence to the MD in 4671 adult subjects from Dalmatia, Croatia (age range 18–98 years; 61.9% were women). Additionally, in the follow-up we examined the change in adherence to the MD and in BMI (subsample, N = 1342; 62.5% were women; mean follow-up time of 5.8 years). The adherence to the MD was based on the Mediterranean Diet Serving Score (range 0–24 points, cut-off value $\geq$ 14 points), with a prevalence in the overall sample of 28.5%. Higher odds of adherence to the MD were recorded in women, older subjects, and those with higher level of objective material status, while it was less likely in the period after economic crisis of 2007–2008. Additionally, we detected no change in adherence to the MD in the follow-up subsample (−8.5%, $p = 0.056$), but there was an increase in BMI (+6.5%, $p < 0.001$). We recorded an increase in adherence for nuts (+127.5%), sweets (+112.6%), red meat (+56.4%), and wine (+50.0%), unlike the reduction in adherence for vegetables (−35.1%), fish (−23.4%), white meat (−11.6%), cereals (−10.9%), and dairy products (−9.6%). Similar results were obtained across all quartiles of objective material status. Over time, the absolute change in the MD score was positively associated with female gender, age, higher education, and moderate physical activity, but it was negatively associated with adherence to the MD at baseline. BMI change was positively associated with female gender, and negatively with initial BMI, initial adherence to the MD, and MD change. Our findings point towards a less than ideal adherence to the MD in the general population of southern Croatia, and identify important characteristics associated with adherence change over time, informing necessary interventions aimed at increasing MD uptake.

Keywords: Mediterranean diet; adherence; BMI; socioeconomic status

1. Introduction

Unhealthy lifestyle and unhealthy diet in particular are among the foremost public health challenges, with as many as 11 million deaths globally being attributable to suboptimal diet in 2017 [1]. The leading global dietary risk factors for death and disability were high sodium intake, low intake of whole grains, and low intake of fruits [1]. These are highly preventable risk factors that could be addressed by adopting scientifically proven healthy diets at the population level.

One model of healthy eating that is particularly well described in the literature is the Mediterranean diet (MD), which is especially healthful compared to a more westernized dietary pattern [2]. MD is characterized by a high intake of plant-based foods, such as daily intake of vegetables, fruit, whole grains, olive oil, nuts and seeds, and weekly intake of dairy, fish and legumes, alongside frugal use of meat, eggs and sweets [3]. There is a large body of evidence showing that adherence to the MD can preserve human health, with the extra bonus of ensuring environmental sustainability [4,5]. The health benefits of MD are numerous [6,7]. The most important positive effects include reduced all-cause mortality [6,8,9], primary prevention of cardiovascular disease [10], lower cancer incidence and mortality [6,11], reduced risk for development of type II diabetes [12–14], obesity and metabolic syndrome [13]. Benefits of MD also include safeguarding of mental health, such as better cognitive performance with higher adherence to the MD [15], reduced risk of depression and cognitive impairment [16], lesser mental distress [17], and overall better health-related quality of life [18,19]. Moreover, MD was even shown to be an efficient treatment strategy for major depressive episodes [20].

Regardless of these and other health benefits of MD and other traditional diets, global nutrition transition caused by modernization and increased incomes has resulted in deviation from traditional plant-based diets towards higher intake of animal-source food, added sugar and vegetable oils [21]. An analysis of the supply of the most important food components of the traditional MD in several Mediterranean countries has revealed that these countries have experienced a process of Westernization during the period from 1961 to 2001, which was especially pronounced in the European countries of the Mediterranean basin [22].

Major constitutional components of the MD, such as fruit, vegetables, olive oil and fish are still present within the dietary pattern, but the discrepancies between Mediterranean countries and regions have started to emerge more consistently [23]. For example, MD decline was observed in Malta, unlike Sardinia, which was accredited to "modernity and improved living conditions, enhanced commercial availability and increased diversity of food preparation" [24]. However, an overall declining trend in adherence to the MD has been previously demonstrated in many Mediterranean countries [25–27], especially in younger generations [28–33]. On the other hand, some countries have experienced an increase in adherence to the MD among adolescents, such as Israel, where increased consumption of fruits, vegetables, cereals, dairy products, and decreased negative eating behaviors were recorded in 2016 compared to 2003 [34].

Besides the greater convenience of a diet relying on processed foods and ready-to-eat fast food, saving time and effort, these foods are also readily available in our modern urbanized environments. They are appetizing and tasty, and they may be cheaper than whole foods. Indeed, the question of a monetary cost behind the Mediterranean dietary pattern has been previously investigated. Some of these previous studies have shown that greater adherence to the MD was associated with a higher dietary cost [35–37], especially if it is compared to a Western dietary pattern [38]. Therefore, it is not surprising to consistently find that the lowest-income households had the lowest adherence to the MD and the highest obesity prevalence [39]. However, it was shown that a higher educational status could exhibit a mitigating effect on poorer diet in lower income countries [40]. These findings demonstrate a complex interplay between different socio-economic determinants and dietary habits.

Furthermore, since we can define socio-economic status (SES) by using several characteristics, it may be challenging to disentangle the main SES contributor to various health outcomes. SES characteristics include objective indicators, such as attained level of education, profession, employment/unemployment status, income, and the subjective perception of one's wealth compared to other people within the same community. Despite this complexity, the impact of SES on dietary pattern is undeniably important. This effect was summarized nicely in a recent paper stating that people "who are better off consume healthier diets than those less well-to-do" [41]. Unfortunately, a clear link between low SES, poor health and obesity was also recognized [41], making it a double priority in terms of the need for effective public health interventions and more broader political, economic and societal interventions against inequalities. In this context, the MD and the overall Mediterranean lifestyle could lend itself "as the most appropriate regime for disease prevention, a sort of complete lifestyle plan for the pursuit of healthcare sustainability" [41]. Indeed, it was consistently shown that people more adherent to the MD had more favorable anthropometric indicators. For example, a large cohort study with a mean of 12 years of follow-up showed that people with high adherence to the MD had a lower risk of becoming overweight/obese, experienced lesser 5-year change in waist circumference, and had lower 5-year weight change in the case of normal weight at baseline [42]. Additionally, MD was found to be more effective in long-term weight loss (over two years of follow-up) in patients with metabolic syndrome than a prudent control diet [43]. It was also found that in older Mediterranean individuals with excess weight, those subjects who desired higher weight loss actually had lower adherence to the MD and higher prevalence of obesity [44]. Hence, MD could serve as a good model for both keeping weight stable across life, and for sustainable weight loss [45].

There is a paucity of studies investigating the trend in adherence to the MD in Croatia. In general, based on geographical location and cultural heritage, the population of the Adriatic region of Croatia is adherent to the MD and the Mediterranean lifestyle [46]. Additionally, Croatia was one of the countries that supported in the inclusion of MD on the UNESCO's Representative List of the Intangible Cultural Heritage of Humanity [47]. However, the role of different socio-economic characteristics in the MD pattern and BMI change in Croatia has been only marginally investigated. It was previously shown that a lower education level was associated with lower adherence to the MD in the population of southern Croatia, while the overall prevalence of adherence to the MD was also rather low [31]. On the other hand, Croatia is heavily encumbered with non-communicable diseases [48], and ranks high among the leading countries in Europe regarding the prevalence of overweight and obesity, with 58% of the adult population being affected [49]. This undesirable trend is present even in young children, with as many as 35.9% of 7–9 year-olds being overweight or obese [50]. Therefore, our aim was to estimate the temporal trend in adherence to the MD and the contribution of several socio-economic factors in the changing pattern of the MD and BMI in a follow-up study including a large sample from Dalmatia, Croatia.

2. Materials and Methods

2.1. Study Participants

This study included 4988 subjects, between 18 and 98 years old, from several settlements in Dalmatia, Croatia, upon their initial enrolment within the "10,001 Dalmatians" study [51], while the follow-up data were available for 1342 subjects. The main objective of the "10,001 Dalmatians" study was to explore genetic and environmental risk factors by creating a biobank in the isolated populations of the Adriatic islands.

Chronologically, the initial field study was performed during 2003 and 2004 on the Island of Vis (N = 1029). An additional 969 subjects were enrolled from the Island of Korčula in 2007 (the Town of Korčula and surrounding settlements), followed by 1012 subjects from the City of Split in 2008–2009. Finally, 857 subjects were included in 2013 from the villages of Smokvica and Čara, situated in the central part of the Island of Korčula, and 1121 subjects were included during 2014–2015 from the towns of Blato and Vela Luka on the western part of the Island of Korčula.

The initial population-based convenient sampling approach employed personal invitations by general practitioners, postal invitations, local media and support from other local stakeholders, namely local governments and priests. Only subjects older than 18 were eligible to participate in the study, without any other restrictions or exclusion criteria.

After being formally informed of the study objectives, subjects signed the informed consent before the enrolment.

The field-based follow-up data collection was performed in 2011 for the subjects from the Island of Vis (N = 482, response rate 46.8%, mean follow-up of 7.5 years). In 2013 we collected follow-up data for the subjects from the Town of Korčula who were initially included in 2007 (N = 366; 37.8%; mean follow-up of 5.3 years), and in 2012–2013 for the subjects from the City of Split (N = 494; 48.8%, mean follow-up of 4.4 years). The main reason for the different follow-up times between study sites is the use of an open cohort sampling approach; this inevitably led to a different amount of time that each participant could be followed for. Subjects from Smokvica, Čara, Blato and Vela Luka (N = 1978) were not included in the follow-up due to their initial inclusion in 2013–2015, after which no additional data collections were done within the "10,001 Dalmatians" study.

The study was approved by the Ethical Committee of the University of Split School of Medicine.

2.2. *Data Collection and Measurements*

Trained nurses and medical doctors performed anthropometric measurements and collected clinically relevant information using the standard operating procedures at the newly established study site in each location. Individual medical histories were taken, together with an extensive self-administered questionnaire (including demographic characteristics, detailed socioeconomic status, dietary habits, smoking habits, alcohol consumption, and physical activity). Elderly people and those with any disabilities were offered assistance during surveying by a team of nine trained surveyors.

Medical records or subjects' responses were used to extract relevant medical history information, including previous diagnoses and the usage of medications for hypertension, diabetes, coronary heart disease (CHD), cerebrovascular insult (CVI), cancer, bipolar disorder, hyperlipidemia and gout.

2.3. *Socioeconomic Status*

Socioeconomic status was assessed during the initial data collection using three determinants: education, subjective material status, and objective material status. Education was categorized into three groups in order to correspond to the Croatian educational system [52]. The three groups were constructed according to the number of completed years of schooling, which corresponded to primary education ($\leq$8 years of schooling), secondary (high school level with 9–12 years of schooling), and higher education ($\geq$13 years). Only 17 subjects reported being students during the initial data collection, and they were automatically included in the higher education group of education.

Subjective material status was assessed based on the participant's perception of her/his material status in comparison to other people in their community. Possible responses on this question were 'much worse than the average', 'somewhat worse than the average', 'the same as others', 'better than the average', 'much better than the average'. These responses were grouped into three categories for easier interpretation: worse than average (responses 'much worse than average' and 'somewhat worse than average'), average ('the same as others'), and better than average (including answers 'better than average' and 'much better than average').

Assessment of objective material status was obtained based on the possession of 16 material items or goods, including heating system, wooden floors, video/DVD recorder, telephone, computer, two TVs, freezer, dishwasher, water supply system, flushing toilet, bathroom, library with more than 100 books, paintings or other art, a car, vacation house or second apartment, and boat, as in our previous study [53]. The sum of those items in the subject's possession indicated the wealth of the subject. Based on the distribution of these wealth scores, quartiles of objective material status were formed: the first quartile with values $\leq$ 8, second quartile with 9–10, third quartile with 11–12, and fourth quartile with values 13–16, as in our previous study [52].

Formal income was not taken into account due to the long period of observation included in this study (from 2003–2015), during which many economic and social changes happened in Croatia, including the financial crisis of 2007–2008. In order to take this into account in our analysis, we have introduced the variable for the recession period (before/after), denoting it as having started in our target population after 2008 (and including subsequent years).

2.4. Mediterranean Diet Assessment

Assessment of the Mediterranean dietary pattern was based on the food frequency questionnaire (FFQ), which was adjusted for application in the population of Dalmatia. There were 55 questions on commonly consumed foods, with 6 possible responses (every day, 2–3 times a week, once a week, once a month, rarely, and never), investigating the frequency of consumption of olive oil and other fats, milk and dairy products, vegetables, fruits, nuts, legumes, various meats, fish and sea foods, eggs, sweets, potatoes, rice, pasta, and bread [31,52]. Mediterranean diet adherence was assessed using the Mediterranean Diet Serving Score (MDSS), which incorporates 14 typical food groups representing the modern MD pyramid: fruit, vegetables, cereals, potatoes, olive oil, nuts, dairy products, legumes, eggs, fish, white meat, red meat, sweets, and fermented beverages—namely wine [54]. MDSS and adherence to the MD were calculated as described previously [31,52], and subjects were classified as adherent to the MD in case they had reached ≥14 points (the range was 0–24 points, with no negative points). MDSS requires a daily intake of vegetables, fruit, olive oil, and cereals (intake of each group is awarded with three points for two or more servings a day). Daily intake is encouraged for nuts and dairy products (each group is awarded with two points for one or more servings a day), and for wine (one or two glasses per day, awarded with one point) [54]. The remaining food groups are awarded with one point. Namely, red meat and sweets should be among the less frequently eaten foods (two or less servings per week), while potatoes, legumes, eggs, fish, and white meat should be consumed weekly. This questionnaire was also validated for use in the Croatian population in the short form [55].

We have excluded 317 subjects from the analysis due to missing values in the FFQ and the inability to calculate the MDSS at baseline.

2.5. Lifestyle Characteristics

Besides diet and socioeconomic factors, we assessed other lifestyle indicators, such as smoking and physical activity. According to smoking status, we divided subjects into current smokers, ex-smokers (those who reported they ceased smoking more than a year ago), and those who had never smoked. Assessment of physical activity included activity during both the working part of the day and the leisure part of the day. Those subjects who reported hard intensity labour or other high-intensity activity during either part of the day were considered as intensively physically active. Subjects who reported moderate intensity of physical activity in either part of the day were considered moderately active, while all others reporting either sitting or light physical activity in both parts of the day were considered as having light physical activity.

Additionally, body mass index (BMI) was calculated using measured height and weight. BMI was divided into three categories, representing subjects with normal body weight (from 18.5 to 24.9 kg/m^2), overweight (25.0 to 29.9 kg/m^2), and obese subjects ($\geq$30.00 kg/m^2).

2.6. Statistical Analysis

All categorical variables were described using absolute numbers and percentages. All numerical variables were described using median and interquartile range (IQR), due to non-normal distribution, which was tested using the Kolmogorov–Smirnov test. The χ^2 test was used to examine the differences between groups for categorical variables and Kruskal–Wallis for numerical variables. We additionally investigated the differences between included subgroups; Mann–Whitney U test was used for pairwise comparison of numerical variables and χ^2 for categorical variables.

Univariate and multivariate logistic regression analysis (enter method) were used to assess the association between three SES characteristics (education level, subjective material status, objective material status) and overall adherence to the MD (MDSS $\geq$ 14 points) at baseline. Additionally, multivariate logistic regression analysis was used for assessing predictors for adherence for each of the 14 MD food groups within the MDSS scoring system. All multivariate models included age, sex, place of residence, number of chronic diseases diagnosed previously, smoking, physical activity, and BMI as confounding factors. There were only 53 subjects in the baseline sample with BMI less than 18.5 kg/m^2, and we have excluded them from the regression analysis due to the small sample size of the group. Additionally, in order to control for the potential confounding effects of the recession of 2007–2008, we included a variable denoting the time period of data collection as being either before or after the recession period in all of the regression models. All of the included covariates were entered as categorical variables to enable easier interpretation of the results. Odds ratios (OR) and 95% confidence intervals (CI) were provided for both univariate and multivariate logistic regression models. Correlations between the three variables describing socioeconomic status were tested using the Spearman rank test, before using them together in logistic regression models; none of the Spearman's rho values were higher than 0.401.

Linear regression models were used to assess the association between absolute change in MDSS and BMI across the follow-up period with different subjects' characteristics. The main predictor variables were again the three SES characteristics (education level, subjective material status, and objective material status), and the models also included important confounding variables: age, follow-up time, sex, place of residence, number of chronic diseases diagnosed previously, smoking, physical activity, BMI at baseline, and MDSS at baseline. Additionally, the model with BMI change during the follow-up as an outcome variable also included the MDSS absolute change during the follow-up as a covariate.

The change in the prevalence of the adherence to the MD and each of the MDSS food groups between baseline (t_0) and the follow-up time period (t_1) was assessed by calculating the percent change, using the following formula:

$$MD\ adherence\ (\%)_{change} = \frac{\text{MD adherence}(\%)_{t_1} - \text{MD adherence}(\%)_{t_0}}{\text{MD adherence}(\%)_{t_0}} * 100, \quad (1)$$

Additionally, the absolute change in MDSS score and BMI between baseline (t_0) and the follow-up time period (t_1) was calculated using the following formulas:

$$MDSS_{change} = \text{MDSS}_{t_1} - \text{MDSS}_{t_0} \quad (2)$$

$$BMI_{change} = \text{BMI}_{t_1} - \text{BMI}_{t_0} \quad (3)$$

The Wilcoxon signed-rank test and McNemar test were used to compare the differences between paired data for repeated measurements (baseline vs. follow-up).

The significance level was set at $p < 0.05$. All statistical analyses were carried out using IBM SPSS Statistics v21 (IBM, Armonk, NY, USA).

3. Results

The analysis included 4671 subjects in total (Table 1). Subjects from the Island of Vis were on average older, less educated, and had the highest average BMI (median of 27.08; IQR 6.05). The median MD adherence score (MDSS) was the lowest in subjects from the Island of Korcula (11 out of 24 points; IQR 6), and it was slightly higher in both subjects from the City of Split and the Island of Vis (median 12; IQR 5). Significant differences in median MDSS score were also recorded between settlements and according to age groups (Table 1). A wide range of adherence to MDSS components was present, ranging from as low as 2.7% for nuts in subjects from Vis, and up to 97.4% adherence for cereals in the same group (Table 1).

Table 1. Demographic characteristics and adherence to the MD (14 food components and overall adherence expressed as MDSS ≥ 14 points), according to the place of residence in a total sample of 4671 subjects.

	Island of Vis N = 1012	Island of Korčula N = 2651	City of Split N = 1008	Overall p (Pairwise Comparison p Values)
Sex; n (%)				
Men	423 (41.8)	967 (36.5)	391 (38.8)	0.011 (0.003 V-K, 0.168 V-S, 0.196 K-S)
Women	589 (58.2)	1684 (63.5)	617 (61.2)	
Age (years); median (IQR)	56.00 (24.00)	55.00 (23.25)	52.00 (21.00)	<0.001 (<0.001 V-K, <0.001 V-S, <0.001 K-S)
Education (years of schooling); median (IQR)	11.00 (4.00)	12.00 (3.00)	12.00 (4.00)	<0.001 (<0.001 V-K, <0.001 V-S, <0.001 K-S)
Subjective material status; median (IQR)	3.00 (0.00)	3.00 (1.00)	3.00 (1.00)	<0.001 (<0.001 V-K, <0.001 V-S, <0.001 K-S)
Objective material status; median (IQR)	10.00 (5.00)	10.00 (3.00)	12.00 (3.00)	<0.001 (<0.001 V-K, <0.001 V-S, <0.001 K-S)
Body mass index (kg/m^2); median (IQR)	27.08 (6.05)	24.59 (5.94)	26.60 (5.63)	<0.001 (<0.001 V-K, 0.024 V-S, <0.001 K-S)
Chronic diseases *; n (%)				
None	542 (53.6)	1565 (59.0)	677 (67.2)	<0.001 (0.011 V-K, <0.001 V-S, <0.001 K-S)
1	289 (28.6)	677 (25.5)	248 (24.6)	
≥2	181 (17.9)	409 (15.4)	83 (8.2)	
Smoking (pack years); median (IQR)	0.00 (10.00)	0.00 (3.00)	0.00 (3.00)	0.004 (0.002 V-K, 0.007 V-S, 0.804 K-S)
Smoking; n (%)				
current smokers	288 (28.5)	741 (28.0)	266 (26.5)	<0.001 (0.001 V-K, 0.105 V-S, 0.005 K-S)
ex-smokers	303 (30.0)	584 (22.2)	275 (27.4)	
never-smokers	419 (41.5)	1306 (49.6)	464 (46.2)	
Physical activity; n (%)				
light	264 (26.2)	537 (20.5)	358 (35.6)	<0.001 (<0.001 V-K, <0.001 V-S, <0.001 K-S)
moderate	580 (57.5)	1815 (69.2)	610 (60.7)	
intensive	164 (16.3)	271 (10.3)	37 (3.7)	
MDSS; median (IQR)	12.00 (5.00)	11.00 (6.00)	12.00 (5.00)	<0.001 (<0.001 V-K, 0.554 V-S, 0.001 K-S)
MDSS according to age group; median (IQR)				
18.0–34.9	10.00 (5.00)	9.00 (5.00)	10.00 (5.00)	0.009 (0.004 V-K, 0.225 V-S, 0.068 K-S)
35.0–64.9	12.00 (5.00)	11.00 (6.00)	12.00 (5.00)	0.001 (0.004 V-K, 0.982 V-S, 0.003 K-S)
≥65.0	12.00 (5.00)	12.00 (6.00)	13.00 (5.00)	<0.001 (0.009 V-K, <0.001 V-S, <0.001 K-S)

Table 1. *Cont.*

	Island of Vis N = 1012	Island of Korčula N = 2651	City of Split N = 1008	Overall p (Pairwise Comparison p Values)
MDSS components adherence; n (%)				
fruit	596 (58.9)	1399 (52.8)	636 (63.1)	<0.001 (0.001 $^{V\text{-}K}$, 0.053 $^{V\text{-}S}$, <0.001 $^{K\text{-}S}$)
vegetables	439 (43.4)	980 (37.0)	418 (41.5)	0.001 (<0.001 $^{V\text{-}K}$, 0.385 $^{V\text{-}S}$, 0.012 $^{K\text{-}S}$)
cereals	986 (97.4)	2367 (89.3)	929 (92.2)	<0.001 (<0.001 $^{V\text{-}K}$, <0.001 $^{V\text{-}S}$, 0.009 $^{K\text{-}S}$)
olive oil	586 (57.9)	1835 (69.2)	643 (63.8)	<0.001 (<0.001 $^{V\text{-}K}$, 0.007 $^{V\text{-}S}$, 0.002 $^{K\text{-}S}$)
nuts	27 (2.7)	117 (4.4)	71 (7.0)	<0.001 (0.015 $^{V\text{-}K}$, <0.001 $^{V\text{-}S}$, 0.001 $^{K\text{-}S}$)
dairy products	256 (25.3)	592 (22.3)	270 (26.8)	0.010 (0.057 $^{V\text{-}K}$, 0.446 $^{V\text{-}S}$, 0.005 $^{K\text{-}S}$)
potatoes	686 (67.8)	1774 (66.9)	823 (81.6)	<0.001 (0.617 $^{V\text{-}K}$, <0.001 $^{V\text{-}S}$, <0.001 $^{K\text{-}S}$)
legumes	326 (32.2)	714 (26.9)	252 (25.0)	0.001 (0.002 $^{V\text{-}K}$, <0.001 $^{V\text{-}S}$, 0.236 $^{K\text{-}S}$)
eggs	297 (29.3)	662 (25.0)	246 (24.4)	0.013 (0.007 $^{V\text{-}K}$, 0.012 $^{V\text{-}S}$, 0.723 $^{K\text{-}S}$)
fish	838 (82.8)	1769 (66.7)	692 (68.7)	<0.001 (<0.001 $^{V\text{-}K}$, <0.001 $^{V\text{-}S}$, 0.269 $^{K\text{-}S}$)
white meat	499 (49.3)	1077 (40.6)	381 (37.8)	<0.001 (<0.001 $^{V\text{-}K}$, <0.001 $^{V\text{-}S}$, 0.118 $^{K\text{-}S}$)
red meat	261 (25.8)	700 (26.4)	248 (24.6)	0.537 (0.705 $^{V\text{-}K}$, 0.539 $^{V\text{-}S}$, 0.266 $^{K\text{-}S}$)
sweets	181 (17.9)	808 (30.5)	168 (16.7)	<0.001 (<0.001 $^{V\text{-}K}$, 0.469 $^{V\text{-}S}$, <0.001 $^{K\text{-}S}$)
wine	204 (20.2)	459 (17.3)	177 (17.6)	0.124 (0.046 $^{V\text{-}K}$, 0.136 $^{V\text{-}S}$, 0.861 $^{K\text{-}S}$)
Adherence to the MD (MDSS $\geq$ 14 points); n (%)	315 (31.1)	711 (26.8)	306 (30.4)	0.012 (0.009 $^{V\text{-}K}$, 0.708 $^{V\text{-}S}$, 0.033 $^{K\text{-}S}$)
Adherence to the MD according to age group (MDSS $\geq$ 14 points); n (%)				
18.0–34.9 years	22 (20.0)	49 (12.3)	26 (14.8)	0.012 (0.026 $^{V\text{-}K}$, 0.745 $^{V\text{-}S}$, 0.070 $^{K\text{-}S}$)
35.0–64.9 years	158 (29.3)	393 (25.4)	203 (30.5)	
$\geq$65.0 years	135 (37.2)	269 (38.4)	77 (46.1)	

IQR—interquartile range; MDSS—Mediterranean Diet Serving Score; MD—Mediterranean diet; p values for categorical variables were obtained with the chi-squared test, and for numerical variables with the Kruskal–Wallis test. Pairwise comparison p values for categorical variables were obtained with the chi-squared test, and for numerical variables with Mann–Whitney U test. * chronic diseases included any or more than one of the following diagnoses: hypertension, diabetes, CHD, CVI, cancer, bipolar disorder, hyperlipidemia and gout. $^{V\text{-}K}$ Pairwise comparison p value: Island of Vis vs. Island of Korčula. $^{V\text{-}S}$ Pairwise comparison p value: Island of Vis vs. City of Split. $^{K\text{-}S}$ Pairwise comparison p value: Island of Korčula vs. City of Split.

Less than half of all of the subjects were compliant with the daily requirement for vegetable intake (lowest on Korčula; 37.0%), while it was a little better for intake of fruit (lowest on Korčula; 52.8%), and olive oil (lowest on Vis; 57.9%). Only 22.3% of subjects from the Island of Korčula, 25.3% from the Island of Vis and 26.8% from the City of Split adhered to the daily dairy products consumption requirement, which was similar for wine (17.3–20.2%). Consistently, the best adherence was recorded for cereals, and the lowest for nuts (Table 1). A total of 1332 subjects (28.5%) were considered as being adherent to the MD pattern in the overall sample. The lowest prevalence was recorded for subjects from the Island of Korčula (26.8%), followed by those from the City of Split (30.4%), and the Island of Vis (31.1%). There was a significant difference in the prevalence of adherence to the MD according to age groups and place of residence, and a significant result was obtained for the comparison between subjects from Vis and Korčula ($p = 0.026$; Table 1).

Logistic regression analysis revealed several characteristics that were strongly associated with adherence to the MD throughout the entire sample (Table 2). Women presented higher odds of adherence compared to men (OR = 1.85, 95% CI 1.58–2.17, $p < 0.001$), while the oldest age group had 3.81-fold higher odds of adherence compared to the youngest subjects (95% CI 2.83–5.12, $p < 0.001$; Table 2). In the fully adjusted model, subjects from the Island of Korčula presented with higher odds of adherence compared to the subjects from the City of Split (OR = 1.63, 95% CI 1.31–2.102, $p < 0.001$). Education level and subjective material status were not associated with adherence to the MD in the adjusted model, unlike objective material status. The wealthiest subjects according to the objective material status (those in the fourth quartile of distribution) were almost twice as likely to be adherent to the MD, compared to subjects in the lowest quartile (OR = 1.93, 95% CI 1.53–2.43, $p < 0.001$). Subjects in the second and third quartile of objective material status also had greater odds of being adherent to the MD (Table 2).

Table 2. Characteristics associated with adherence to the MD (MDSS $\geq$ 14 points) in the total sample (N = 4671), as determined by the logistic regression analysis.

		Unadjusted Odds Ratio (95% Confidence Interval); *p*	**Adjusted Odds Ratio (95% Confidence Interval); *p***
Sex			
	Male; Ref.	1.00	1.00
	Female	1.60 (1.39, 1.83); <0.001	1.85 (1.58, 2.17); <0.001
Age group			
	18–34.9; Ref.	1.00	1.00
	35–64.9	2.29 (1.82, 2.88); <0.001	1.99 (1.54, 2.57); <0.001
	≥65.0	3.89 (3.05, 4.97); <0.001	3.81 (2.83, 5.12); <0.001
Place of residence			
	City of Split; Ref.	1.00	1.00
	Island of Vis	1.04 (0.86, 1.25); 0.708	1.04 (0.84, 1.29); 0.696
	Island of Korčula	0.84 (0.72, 0.99); 0.033	1.63 (1.31, 2.02); <0.001
Education (Years of schooling)			
	elementary (0–8); Ref.	1.00	1.00
	high school (9–12)	0.69 (0.78, 0.80); <0.001	0.93 (0.77, 1.14); 0.492
	higher (13+)	0.94 (0.79, 1.12); 0.494	1.19 (0.95, 1.5); 0.130
Subjective material status			
	worse than average; Ref.	1.00	1.00
	average	1.13 (0.92, 1.39); 0.250	1.14 (0.91, 1.44); 0.258
	better than average	1.28 (1.03, 1.61); 0.028	1.16 (0.89, 1.51); 0.267

Table 2. *Cont.*

	Unadjusted Odds Ratio (95% Confidence Interval); *p*	Adjusted Odds Ratio (95% Confidence Interval); *p*
Objective material status		
1st quartile; Ref.	1.00	1.00
2nd quartile	1.12 (0.94, 1.35); 0.216	1.38 (1.12, 1.70); 0.002
3rd quartile	0.98 (0.82, 1.17); 0.791	1.29 (1.04, 1.61); 0.020
4th quartile	1.52 (1.27, 1.83); <0.001	1.93 (1.53, 2.43); <0.001
Chronic diseases *		
≥2; Ref.	1.00	1.00
1	0.85 (0.69, 1.04); 0.107	0.93 (0.75, 1.17); 0.546
none	0.69 (0.57, 0.82); <0.001	0.93 (0.75, 1.16); 0.507
Smoking		
current smokers; Ref.	1.00	1.00
ex-smokers	1.70 (1.41, 2.03); <0.001	1.40 (1.14, 1.71); 0.001
never-smokers	1.75 (1.49, 2.06); <0.001	1.36 (1.13, 1.63); 0.001
Physical activity		
light; Ref.	1.00	1.00
moderate	1.24 (1.07, 1.45); 0.005	1.44 (1.21, 1.70); <0.001
intensive	1.16 (0.91, 1.48); 0.222	1.50 (1.15, 1.97); 0.003
Body mass index category #		
18.0–24.9 (kg/m^2); Ref.	1.00	1.00
25.0–29.9 (kg/m^2)	0.92 (0.79, 1.05); 0.218	0.98 (0.83, 1.16); 0.834
≥30.0 (kg/m^2)	0.79 (0.66, 0.95); 0.013	0.84 (0.68, 1.05); 0.123
The economic crisis of 2007–2008		
before; Ref.	1.00	1.00
after	0.40 (0.35, 0.46); <0.001	0.31 (0.25, 0.38); <0.001

Adjusted odds ratios, 95% confidence intervals and *p* values were calculated using a multivariate logistic regression model simultaneously adjusted for all the covariates listed in this table (enter method). * chronic diseases included any or more than one of the following diagnoses: hypertension, diabetes, CHD, CVI, cancer, bipolar disorder, hyperlipidemia and gout; # 53 subjects with BMI < 18.5 kg/m^2 were excluded from the analysis due to small sample size of the group and negative impact on the model performance.

Subjects who never smoked and ex-smokers presented with higher odds of adherence to the MD, compared to current smokers (OR = 1.36, 95% CI 1.13–1.63, $p = 0.001$; OR = 1.40, 95% CI 1.14–1.71, $p = 0.001$, respectively). Subjects with higher levels of physical activity were also more likely to be adherent to the MD (Table 2). BMI and diagnosis of chronic diseases were not associated with adherence to the MD. The study period was statistically significantly associated with adherence to the MD, in a way that MD adherence was less likely in the period after the recession (OR = 0.31, 95% CI 0.25–0.38, $p < 0.001$; Table 2).

The fully adjusted regression model yielded a good data fit (Hosmer and Lemeshow $p = 0.304$; Nagelkerke R2 = 0.100).

Determinants of adherence to MD food components are shown in Supplemental Table S1. Women were more likely to be adherent to the recommended intake of fruit, vegetables, olive oil, nuts, diary, and red meat, but they were less likely to be adherent to the eggs and wine intake MD recommendations compared to men (women most commonly abstained from alcohol intake). Older subjects had higher odds for meeting the recommendations for fruit, vegetables, cereals, olive oil, nuts, fish, red meat, sweets, and wine intake, but lower odds for potatoes and eggs adherence compared to the youngest group of subjects. The highest level of education was associated with lesser adherence to the MD guidelines for intake of cereals, olive oil, legumes, fish, and white meat, in contrast to a higher adherence to appropriate intake of dairy products, potatoes and red meat compared to subjects with the lowest level of education (Supplemental Table S1). Subjective material status was less associated with MD food components intake, unlike the objective material status.

Compared to subjects in the lowest quartile of objective material status, subjects belonging to higher quartiles presented with an increasing trend of compliance with fruit, vegetables, olive oil, and fish intake recommendations, but also with a decreasing compliance for the intake of red meat and sweets (Supplemental Table S1).

Obese subjects (BMI 30 $\geq$ kg/m^2) were 34% more likely to adhere to recommendations for sweets, but also 30% less likely to adhere to recommendations for cereals intake, and 42% less adherent for nuts.

The study period after the recession was associated with 68% decreased odds for adherence to vegetables intake recommendations, 55% decreased odds for cereals adherence, 50% for fruit, 49% for fish, 47% for legumes, 36% for dairy products, 31% for potatoes, and 29% decreased odds for adherence to olive oil intake. On the other hand, we recorded 39% increased odds for adherence to red meat and 23% increased adherence to sweets intake recommendation after recession (Supplemental Table S1).

In order to assess the change in Mediterranean diet compliance over time, 1342 subjects were included in the follow-up study. A breakdown by four quartiles of the objective material status demonstrated significant changes in adherence for several MD food groups across the follow-up period (Figure 1). A distinct pattern of change was recorded, with the most prominent and significant decrease in adherence to the recommended intake of vegetables, followed by a decrease in fish and cereals recommended intake across all quartiles of objective material status (Figure 1). On the other hand, a significant increase in adherence for nuts was reported across all quartiles of material status (corresponding to increased intake), followed by an increase in sweets, potatoes and red meat (decreased intake), wine, legumes, and eggs adherence (increased intake). The exception was adherence to wine, legumes, and eggs recommendations in subjects within the lowest quartile of the objective material status, where these results were not significant. Based on such diverse results in individual MDSS food groups, the overall change in adherence to the MD was insignificant in all of the quartiles of objective material status (Figure 1). A similar result was obtained in the total group of subjects included in the follow-up, with a borderline insignificant decrease in adherence to the MD (by 8.5%; from 36.6% of adherent subjects at study baseline, to 33.5% in the follow-up; $p = 0.056$; Table 3). Furthermore, the highest overall increase in adherence was recorded for nuts (127.5%), and sweets (112.6%), followed by red meat (56.4%), and wine (50.0%). On the other hand, the most significant decrease in adherence was recorded for vegetables (−35.1%), followed by fish (−23.4%), white meat (−11.6%), cereals (−10.9%), and dairy products (−9.6%). At the same time, the average BMI had increased from 25.76 kg/m^2 at baseline of the study to 27.44 kg/m^2 at the follow-up time period ($p < 0.001$).

Linear regression analysis revealed several variables that were significantly associated with the MDSS change during the follow-up period (Table 4). MDSS change was positively associated with female gender (β = 0.41; 95% CI 0.00–0.83; $p = 0.049$), age (β = 0.05; 95% CI 0.03–0.06); $p < 0.001$), highest level of education (β = 0.71; 95% CI 0.07–1.36; $p = 0.031$), and with moderate physical activity (β = 0.72; 95% CI 0.27–1.16; $p = 0.002$). MDSS at baseline displayed a negative association with the MDSS change (β = −0.64; 95% CI −0.70–−0.58; $p < 0.001$), while BMI at baseline, smoking, chronic diseases, place of residence, objective and subjective material status were not associated with the absolute change in the Mediterranean Diet Serving Score. The regression model yielded a good data fit (Durbin–Watson = 1.994; Adjusted R2 = 0.280).

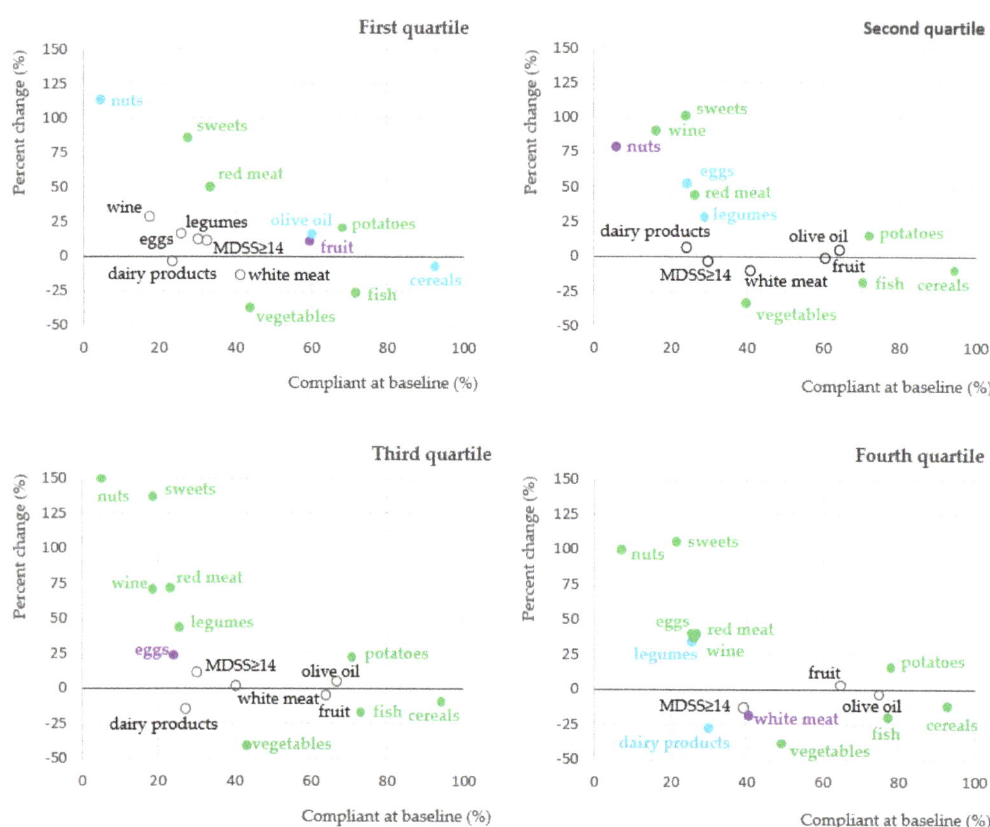

Figure 1. Change in adherence to the MD food components and the overall MD (MDSS $\geq$ 14 points), expressed as a percentage change from baseline to the follow-up, according to the objective material status category. Significant results at the level of $p < 0.05$ are denoted with the full circle (green < 0.001, blue < 0.01, purple < 0.05, McNemar test).

Table 3. Adherence to 14 MD food groups and the overall MD (MDSS $\geq$ 14 points) at baseline and at the follow-up (N = 1342; 366 subjects from Korčula, 494 from Split, and 482 subjects from Vis).

		Baseline N = 1342	Follow Up N = 1342	Percent Change (%)	*p*
Sex; *n* (%)					
	men	503 (37.5)	-	-	-
	women	839 (62.5)			
Age (years); median (IQR)		55.00 (18.00)	62.01 (16.96)	-	-
Age group; *n* (%)					
	18.0–34.99	127 (9.5)	58 (4.3)	-	-
	35.0–64.99	926 (69.0)	724 (53.9)		
	65+	289 (21.5)	560 (41.7)		
Body mass index (kg/m^2); median (IQR)		25.76 (5.74)	27.44 (5.06)	6.5	<0.001
Adherence to the MD (MDSS $\geq$ 14 points); *n* (%)		491 (36.6)	449 (33.5)	−8.5	0.056

Table 3. *Cont.*

	Baseline N = 1342	Follow Up N = 1342	Percent Change (%)	*p*
MDSS components adherence; *n* (%)				
fruit	868 (64.7)	848 (63.2)	−2.3	0.341
vegetables	643 (47.9)	417 (31.1)	−35.1	<0.001
cereals	1277 (95.2)	1138 (84.8)	−10.9	<0.001
potatoes	985 (73.4)	1183 (88.2)	20.2	<0.001
olive oil	893 (66.5)	927 (69.1)	3.9	0.112
nuts	68 (5.1)	156 (11.6)	127.5	<0.001
dairy products	356 (26.5)	309 (23.0)	−9.6	0.030
legumes	400 (29.8)	457 (34.1)	14.4	0.011
eggs	339 (25.3)	405 (30.2)	19.4	0.002
fish	1036 (77.2)	793 (59.1)	−23.4	<0.001
white meat	556 (41.4)	487 (36.3)	−11.6	0.005
red meat	347 (25.9)	544 (40.5)	56.4	<0.001
sweets	276 (20.6)	588 (43.8)	112.6	<0.001
wine	268 (20.0)	403 (30.0)	50.0	<0.001

MDSS—Mediterranean Diet Serving Score. MD—Mediterranean Diet. *p* values for categorical variables were obtained using McNemar test and for numerical using Wilcoxon Signed-Ranks Test.

Table 4. Characteristics associated with the absolute change in the Mediterranean Diet Serving Score (MDSS) and the BMI across the follow-up period, as determined by the linear regression model (sample size is 1342 subjects; all independent variables were included in the model simultaneously).

	MDSS Change during Follow-Up Beta (95% Confidence Interval); *p*	BMI Change during Follow-Up Beta (95% Confidence Interval); *p*
Sex		
Male; Ref.	1.00	1.00
Female	0.41 (0.00, 0.83); 0.049	0.33 (0.06, 0.60); 0.016
Age at baseline (years)	0.05 (0.03, 0.06); <0.001	−0.01 (−0.02, 0.00); 0.192
Follow up time (years)	0.03 (−0.09, 0.16); 0.571	−0.07 (−0.15, 0.01); 0.099
Place of residence		
City of Split; Ref.	1.00	1.00
Island of Vis	0.28 (−0.75, 1.32); 0.590	0.86 (0.18, 1.54); 0.013
Island of Korčula	0.00 (−0.81, 0.81); 0.992	3.68 (3.15, 4.21); <0.001
Education (years of schooling)		
elementary (0–8); Ref.	1.00	1.00
high school (9–12)	−0.11 (−0.67, 0.46); 0.708	−0.05 (−0.42, 0.32); 0.799
higher (13+)	0.71 (0.07, 1.36); 0.031	0.06 (−0.36, 0.48); 0.769
Subjective material status		
worse than average; Ref.	1.00	1.00
average	0.05 (−0.59, 0.69); 0.872	0.15 (−0.26, 0.57); 0.468
better than average	−0.01 (−0.72, 0.71); 0.982	0.11 (−0.36, 0.57); 0.655
Objective material status		
1st quartile; Ref.	1.00	1.00
2nd quartile	−0.14 (−0.71, 0.43); −0.637	0.07 (−0.30, 0.44); 0.724
3rd quartile	0.02 (−0.56, 0.59); 0.955	0.06 (−0.31, 0.44); 0.733
4th quartile	−0.06 (−0.68, 0.56); 0.859	0.02 (−0.38, 0.42); 0.921

Table 4. *Cont.*

	MDSS Change during Follow-Up Beta (95% Confidence Interval); p	BMI Change during Follow-Up Beta (95% Confidence Interval); p
Chronic diseases *		
≥2; Ref.	1.00	1.00
1	−0.23 (−0.91, 0.44); 0.497	0.24 (−0.19, 0.68); 0.276
none	−0.02 (−0.68, 0.64); 0.946	0.26 (−0.17, 0.69); 0.240
Smoking		
current smokers; Ref.	1.00	1.00
ex-smokers	−0.20 (−0.74, 0.34); 0.468	−0.07 (−0.42, 0.28); 0.707
never-smokers	0.12 (−0.38, 0.62); 0.632	0.02 (−0.31, 0.34); 0.909
Physical activity		
light; Ref.	1.00	1.00
moderate	0.72 (0.27, 1.16); 0.002	−0.02 (−0.31, 0.27); 0.887
intensive	0.69 (−0.02, 1.40); 0.057	0.23 (−0.24, 0.70); 0.331
BMI at baseline (kg/m^2)	−0.03 (−0.09, 0.02); 0.188	−0.11 (−0.14, −0.07); <0.001
MDSS at baseline	−0.64 (−0.70, −0.58); <0.001	−0.07 (−0.12, −0.03); 0.001
MDSS change during follow-up	-	−0.04 (−0.07, 0.00); 0.041

* Chronic diseases included any or more than one of the following diagnoses: hypertension, diabetes, CHD, CVI, cancer, bipolar disorder, hyperlipidemia and gout. MDSS—Mediterranean Diet Serving Score.

BMI change during the follow-up period was significantly associated with female gender, place of residence, BMI at baseline, MDSS at baseline and MDSS absolute change (Table 4). Women experienced higher odds for BMI increase compared to men (β = 0.33; 95% CI 0.06–0.60; p = 0.016), the same as subjects from the Island of Vis and Korčula compared to subjects from the City of Split (β = 0.86; 95% CI 0.18–1.54; p = 0.013, and β = 3.68; 95% CI 3.15–4.21; p < 0.001, respectively). BMI at baseline, MDSS at baseline, and MDSS change during the follow-up were all significantly negatively associated with the BMI change (β = −0.11; 95% CI −0.14–−0.07; p < 0.001, β = −0.07; 95% CI −0.12–−0.03; p = 0.001, β = −0.04; 95% CI −0.07–0.00; p = 0.041, respectively), while none of the socio-economic characteristics were associated with absolute BMI change. The regression model yielded good data fit (Durbin–Watson = 1.972; Adjusted R2 = 0.354).

4. Discussion

Our results demonstrated a rather low prevalence of adherence to the MD over the entire sample (28.5%), especially among younger individuals (14.0%). Subjects included in the follow-up had a higher adherence to the MD at baseline (36.6%), with a borderline insignificant decline at the end of the follow-up period (33.5%). On the other hand, BMI had increased on average by 6.5% in subjects available for follow-up.

Our result for MD prevalence was within the expected range, compared to the results from other Mediterranean countries and from Croatia. For example, findings from the literature vary anywhere between 14% of adherent people in Northern Italy [56] to 45% in Balearic Islands [57]. Our current study identified a slightly higher prevalence of adherence to the MD compared to our previous results, when we identified 23% of subjects as adherent to the MD [31]. This difference is due to a smaller sample and different period included in the previous study [31].

Unfortunately, many Mediterranean societies are moving away from their traditional dietary pattern, while some countries in Northern Europe and around the world are adopting a Mediterranean-like dietary pattern [25]. For example, previous studies have indicated a persistent moderate-to-weak adherence to the MD across several southern European countries, including Spain, Portugal, Italy, Greece, and Cyprus [58]. Some variations can be expected, probably due to the applied methodological framework and different instruments used for assessing adherence to the MD [55]. For example, one study from Spain showed a poor level of adherence to the MD in the general population and specific areas of Spain [59], while another one showed moderate adherence [60]. Nevertheless, deflection from a traditional MD diet and lifestyle represents a lost opportunity, not only from the perspective of achieving less-than-ideal individual and population health, but also from the perspective of environmental protection, possible degradation of sociocultural food values, and loss of positive local economic returns [61]. Additionally, a higher prevalence of adherence to the MD in the population can also serve as a safeguard from consumption of ultra-processed foods [62]. This was shown even in very young children from Spain, whose adherence to the traditional MD was inversely associated with energy intake from ultra-processed foods [63].

Previous studies have demonstrated that individual and contextual socio-economic factors are strong determinants of dietary habits and that poorer socio-economic groups are less likely to follow a healthy lifestyle [64]. On the other hand, social position in terms of education, occupational class, and income level represents a good predictor for healthy eating behavior [65,66]. People with a higher educational status have been shown to have a healthier consumption pattern [67]. Higher educational status was also associated with better nutritional intakes in lower GDP countries, while lower-income countries and lower education groups had poorer diets, particularly in terms of micronutrients intake [40].

The current economic and social European context—the increasing crisis, lack of jobs, various challenges due to the COVID-19 pandemic and consequent fall in income associated with cost inflation—could make people inclined to save money in all possible ways. In this context, the most exposed are the disadvantaged groups because they prefer buying food at low prices that are often of low quality [68]. Foods of lower nutritional value and lower-quality diets generally cost less per calorie and tend to be selected by groups of lower socioeconomic status [69]. On the other hand, people with low socio-economic status do not obtain the same health outcomes as those with high socio-economic status, even if both groups follow the same eating pattern [70]. Concretely, high adherence to the MD was associated with cardiovascular protection in higher but not in lower socio-economic groups from Italy, with a similar result observed for both education level and household income groups [70].

In some European countries, it was demonstrated that socio-economic status could modulate adherence to the MD [71,72]. For example, in a study carried out in the adult population from the Balearic Islands, people with a higher educational and socio-economic level showed higher rates of adherence to the Mediterranean pattern [57]. On the other hand, adherence to the MD in the South of Italy was found to be at low levels due to poor knowledge on MD concerning its beneficial effects [73], whereas social status in France was important for healthy eating only through an interaction between level of education and area of residence [64]. A similar association between education and MD was observed in our previous study from Croatia, where less educated people had a reduced likelihood of being adherent to the MD [31]. On the other hand, our current study did not corroborate such a finding, probably due to the inclusion of additional socio-economic indicators in the analysis (subjective and objective material status). Hence, we have identified only a significant association between overall adherence to the MD and objective material status. Subjects reporting the highest objective material status (fourth quartile) demonstrated a 93% higher probability of adhering to the MD than those belonging to the first quartile of objective material status, with similar findings for subjects within the second and third quartile groups (38% and 29%, respectively). This was in line with previous results, where higher household income was positively associated with greater adherence to the MD [39,74].

Interestingly, subjects with a higher educational attainment had a greater probability for appropriate adherence to dairy products, potatoes, and red meat intake recommendations, but they exhibited lesser adherence to cereals, olive oil, legumes, fish, and white meat intake recommendations. This represents a considerable departure from the traditional MD pattern. For example, Biesbroek et al. revealed that people with low education consumed more potatoes, whereas highly educated people consumed more olive oil and fish [75]. Another study showed that highly educated people in Italy also consumed white meat slightly less than in the past [56], while Bonaccio et al. had a similar conclusion for the consumption of white meat, which was again opposite when it came to people with high educational status and consumption of olive oil and fish [70]. Similarly, higher educational status was shown to be positively associated with fish intake [76]. Other MD food components were equally consumed by all educational groups in our sample, as previously shown in another study by Bonaccio et al. [39].

In general, highly educated people have a higher income, and they tend to follow MD recommendations [66]. This could be explained by the fact that greater adherence to the Mediterranean diet was associated with higher dietary cost, which might represent a barrier to healthy eating [35]. For instance, in a study including a representative national sample of 3534 children and young people from Spain, researchers have found that high adherence to the MD was more expensive than low adherence by 0.71 Euros per day [37].

Interestingly, we failed to find any association between subjective material status and adherence to the MD, whereas objective material status presented as the most prominent socio-economic indicator for overall adherence to the MD and for several food groups. For example, subjects in the higher quartiles of material status had higher adherence to fruit, vegetables, olive oil, and fish intake recommendations, but also lower adherence to red meat and sweets intake. Interestingly, olive oil and fish intake had an opposing contribution of educational level and material status to their adherence, such that lower education and higher material status were both associated with greater adherence. This could be explained by the fact that older, less educated people from Dalmatia, especially from remote islands, still tend to produce their own olive oil, and they catch fish on their own, which could be behind their higher intake of these foods (statement based on personal communication with subjects included in the study).

Our results are largely in line with previous studies, which showed that female gender and non-smokers [77–79], older adults [77,80,81], and more physically active people displayed higher adherence to the MD pattern, while higher body mass index was generally associated with lower adherence to the MD [78–80,82]. Our results partially replicated such associations, as subjects with higher levels of physical activity had up to a 50% greater probability of being adherent to the MD in comparison to the ones with light activity, while BMI was not significantly associated with adherence to the MD in our overall sample. This is in contrast to some previous findings [72,83,84], but in line with some studies [85,86]. These differences between previous results are probably due to the employed study design (cross-sectional vs. longitudinal, observational vs. experimental design), and characteristics of included subjects (primarily age and health status), leaving the association between adherence to the MD and BMI a topic for further investigation and open discussion.

The effect of the economic crisis of 2007–2008 on the adherence to the MD was a topic of several previous studies. For instance, it was found that adherence to the MD was lower in subjects from Italy reporting a negative impact of the crisis on their diet [87]. Additionally, the prevalence of adherence to MD among southern Italian citizens enrolled within the Moli-sani study was 31.3% during the 2005–2006 period, which dramatically fell during 2007–2010 (18.3%), most strongly affecting elderly, less affluent people, and urban areas dwellers [88]. Our results also revealed decreased odds for adherence to different food groups, i.e., adherence odds for vegetables, cereals, fruit, fish, legumes, dairy products, potatoes, and olive oil. On the other hand, odds of adherence to red meat and sweets recommendations increased after the recession.

Similar findings were demonstrated in Portugal, where a significant decrease in consumption of fish, fruit and vegetables was recorded from 2005/2006 to 2014 [29]. A cross-sectional study from Greece showed that parents who reported that the financial crisis affected their food spending also reported lower consumption of fruits, carbohydrate foods, and legumes, and increased intake of nutrient-poor/energy-dense foods, while their children had reduced weekly consumption of vegetables and increased weekly consumption of nutrient-poor/energy-dense foods [89]. These and other recent evidence show a possible involvement of the economic crisis, and material resources as strong determinants of adherence to the MD in the period after the recession started [90], given that a direct positive association between the cost of the diet and adherence to the MD has been established [36]. However, it is hard to distinguish the contribution of recession due to the economic crisis from the impact of the steady process of westernization of traditional dietary habits, including MD. For instance, it was noted by FAO that "the Mediterranean region is passing through a 'nutritional transition' in which problems of undernutrition coexist with overweight, obesity and food related chronic diseases" [91]. For example, an ecological study of the changes in food patterns in Europe over the last 40 years revealed that the greatest changes have occurred in Mediterranean Europe [92]. For instance, an increase of 20% in total energy availability was noted, alongside with a 48% increase in energy availability from lipids, and 20% decrease from carbohydrates, with a significant fall in the energy supplied by cereals (30%) and wine (55%), while the contribution of milk and dairy products increased by 78% and 24%, respectively [92]. For example, it was estimated that the Spanish diet shifted away from the traditional MD, now containing three times more meat, dairy and sugar products, and a third fewer fruits, vegetables, and cereals [93]. In our sample available for follow-up, we have detected similar deviations. For example, to our great dismay, vegetables adherence was reduced by 35%, followed by a reduction in fish adherence by 23%, white meat by 12%, cereals by 11%, and dairy products by 10%, while fruit adherence was reduced by only 2%. However, we did record a few positive trends, such as an increase in adherence for nuts (128%), sweets (113%; denoting reduced intake), red meat (56%; also denoting reduced intake), and wine (50%). Overall, the adherence to the MD remained stable, which was probably a consequence of differences in specific MD food constituents.

As already mentioned, a continuous increase in red and processed meat has been observed over the last couple of decades, while fruit, cereals, and vegetable consumption has decreased in different countries [21,94,95]. Our findings are in line with these trends, except for red meat intake, for which compliance was improved. To our satisfaction, an improvement over time was also recorded for sweets adherence in our study, which was in contrast to the findings from Portugal, where sweets/desserts consumption was significantly higher in 2014 compared to 2005/2006 [29]. However, a similar decreased trend for sweets intake were observed in Northern Italy [56], and Norway, Sweden, and Finland [96]. A study conducted among adults in Lebanon showed a decrease in the consumption of bread, fruits, fresh fruit juices, milk and eggs, whereas the consumption of added fats and oils, poultry, cereals and cereal-based products, chips and salty crackers, sweetened milk and hot beverages increased over time [97]. These findings indicate slightly different, yet similar patterns of change in different populations. The important next step in any effort to improve dietary habits in communities is the identification of factors associated with such changes, in order to be able to implement targeted interventions. We have conducted such an analysis, which pointed to several characteristics associated with the change in adherence to the MD over time. Female gender, older age at baseline, the highest level of education, and a moderate level of physical activity were positively associated with MDSS change during follow-up in a multivariate model. On the other hand, the MDSS score at baseline was negatively associated with the MDSS change during the follow-up, indicating that people with a higher baseline adherence to the MD tended to recede over time, while those with lower adherence strived toward increasing adherence to the MD. These findings highlight a continuous change of dietary patterns in the population, requiring constant monitoring of trends and identification of the drivers of such change. This is relevant from the perspective of population health and delivery of adequate health care, as well as from the perspective of economic, social and cultural development.

The importance of a healthy lifestyle and healthy dietary habits came to the frontline of attention due to the COVID-19 pandemic. For example, preliminary findings from the ecological study showed that Mediterranean diet adherence was negatively associated with both COVID-19 cases and related deaths in Spain and 23 other OECD countries, which the authors attributed to the anti-inflammatory properties of the Mediterranean diet [98]. On the other hand, an unhealthy lifestyle and associated metabolic disturbances and concomitant chronic diseases were shown to increase the risk for adverse outcomes after SARS-CoV-2 infection [99].

The traditional Mediterranean diet was shown to be beneficial in the prevention of weight gain and abdominal obesity [42]. On the other hand, a lower educational level was often found to be associated with a higher prevalence of overweight and obesity [39,100]—the same as economic affluence at a country level, reflecting a potential adverse outcome concomitant with economic growth [101]. While the relationship between socio-economic status and health outcomes was frequently emphasized for the Mediterranean area [100], the synergy between those two determinants was not substantially investigated in the population of Southern Croatia. Our results indicate that the average BMI had increased from 25.76 kg/m^2 at baseline to 27.44 kg/m^2 during the follow-up period. This is consistent with the trend of increasing rates of obesity across 147 countries [101]. BMI change during follow-up was positively associated with female gender, and negatively with initial BMI, initial adherence to the MD, and with change in adherence to the MD, as found in the regression analysis. This means that people with a lower BMI at the beginning of the study tended to experience a rise in BMI, while those who started with higher values managed to diminish it over time. An encouraging finding is that individuals with a higher-level/score in MD adherence experienced lower BMI change or even its decrease.

An important limitation of our study that needs to be mentioned here is the use of the cross-sectional design for estimating the association between socio-economic status and adherence to the MD, which limits the inference on causality. However, we did employ an additional follow-up study design in order to confirm the initial findings and observe time trends, as well as to investigate the association between initial socio-economic status and change in adherence to the MD, and BMI change in our sample. Another limitation is the broad sampling period of subjects included in our study, which stretched from 2003 to 2015. In order to control for the effect of the study period in the logistic regression analysis, we included the actual follow-up time as one of the predictor variables, as well as the variable "economic crisis of 2007–2008". We also managed to obtain a smaller than ideal sample size and lesser response rate in the follow-up study (28.7%), due to the older age of subjects and their inability to participate in the follow-up examination. Advantages of the study include a relatively large overall sample size, inclusion of many potential predictors, and sampling from the general population of inhabitants from the Mediterranean region of Croatia. Determinants of Mediterranean diet adherence were so far only marginally investigated in the population of Dalmatia in Croatia, and we are filling this gap.

In conclusion, this is the first study from Croatia to examine the changes in adherence to the MD over time. Additionally, we have identified several important characteristics associated with greater adherence to the MD and with its change over time. These insights should be used to inform the necessary and targeted interventions aimed at increasing MD uptake in order to ensure beneficial outcomes. These include, but are not limited to, the promotion and advancement of individual and population health, ensuring environmental sustainability, and positive impacts on local economies and tourism, as well as the very important outcome of the preservation of cultural heritage for generations to come.

Supplementary Materials: The following are available online at https://www.mdpi.com/article/10.3390/nu13113802/s1, Table S1: Characteristics associated with adherence to the main food groups within the Mediterranean diet serving score (MDSS), as determined by the multivariate logistic regression analysis (N = 4671).

Author Contributions: Conceptualization, I.K., A.P., C.H. and O.P.; data collection, all authors; methodology, I.K., A.P., M.M., C.H. and O.P.; software, I.K. and A.P.; validation, O.P., I.K., R.P., F.P.S. and A.P.; formal analysis, A.P.; investigation, all authors; resources, O.P., C.H. and I.K.; data curation, all authors; writing—original draft preparation, A.P., R.P., F.P.S., M.V., M.M. and I.K.; writing—review and editing, all authors.; visualization, A.P., M.V. and I.K.; supervision, O.P. and I.K.; funding acquisition, I.K., O.P. and C.H. All authors have read and agreed to the published version of the manuscript.

Funding: This study was funded by grants from the Medical Research Council (UK), the European Commission Framework 6 project EUROSPAN (Contract No. LSHG-CT-2006-018947), the European Commission Framework 7 project BBMRI-LPC (FP7 313010), the Republic of Croatia Ministry of Science, Education and Sports research grant (216-1080315-0302), the Croatian Science Foundation (grant 8875), the Croatian National Center of Research Excellence in Personalized Healthcare (grant number KK.01.1.1.01.0010), and the Center of Competence in Molecular Diagnostics (KK.01.2.2.03.0006).

Institutional Review Board Statement: The study was conducted according to the guidelines of the Declaration of Helsinki and approved by the Institutional Ethics Committee of the University of Split School of Medicine (protocol code 2181-198-03-04/10-11-0008).

Informed Consent Statement: Informed written consent was obtained from all subjects involved in the study.

Data Availability Statement: The datasets generated during and/or analyzed during the current study are available from the corresponding author upon reasonable request.

Acknowledgments: We would like to acknowledge the staff of several institutions in Croatia that supported the fieldwork, especially the Institute for Anthropological Research from Zagreb, Croatia.

Conflicts of Interest: The authors declare no conflict of interest.

References

1. Afshin, A.; Sur, P.J.; Fay, K.A.; Cornaby, L.; Ferrara, G.; Salama, J.S.; Mullany, E.C.; Abate, K.H.; Abbafati, C.; Abebe, Z.; et al. Health effects of dietary risks in 195 countries, 1990–2017: A systematic analysis for the Global Burden of Disease Study 2017. *Lancet* **2019**, *393*, 1958–1972. [CrossRef]
2. Trichopoulou, A.; Lagiou, P. Healthy traditional Mediterranean diet: An expression of culture, history, and lifestyle. *Nutr. Rev.* **1997**, *55*, 383–389. [CrossRef]
3. Bach-Faig, A.; Berry, E.M.; Lairon, D.; Reguant, J.; Trichopoulou, A.; Dernini, S.; Medina, F.X.; Battino, M.; Belahsen, R.; Miranda, G.; et al. Mediterranean diet pyramid today. Science and cultural updates. *Public Health Nutr.* **2011**, *14*, 2274–2284. [CrossRef]
4. Berry, E.M. Sustainable Food Systems and the Mediterranean Diet. *Nutrients* **2019**, *11*, 2229. [CrossRef]
5. Dernini, S.; Berry, E.M. Mediterranean Diet: From a Healthy Diet to a Sustainable Dietary Pattern. *Front. Nutr.* **2015**, *2*, 15. [CrossRef]
6. Dinu, M.; Pagliai, G.; Casini, A.; Sofi, F. Mediterranean diet and multiple health outcomes: An umbrella review of meta-analyses of observational studies and randomised trials. *Eur. J. Clin. Nutr.* **2018**, *72*, 30–43. [CrossRef]
7. Dominguez, L.J.; Di Bella, G.; Veronese, N.; Barbagallo, M. Impact of Mediterranean Diet on Chronic Non-Communicable Diseases and Longevity. *Nutrients* **2021**, *13*, 2028. [CrossRef] [PubMed]
8. Trichopoulou, A.; Orfanos, P.; Norat, T.; Bueno-de-Mesquita, B.; Ocke, M.C.; Peeters, P.H.; van der Schouw, Y.T.; Boeing, H.; Hoffmann, K.; Boffetta, P.; et al. Modified Mediterranean diet and survival: EPIC-elderly prospective cohort study. *BMJ* **2005**, *330*, 991. [CrossRef] [PubMed]
9. Trichopoulou, A.; Costacou, T.; Bamia, C.; Trichopoulos, D. Adherence to a Mediterranean diet and survival in a Greek population. *N. Engl. J. Med.* **2003**, *348*, 2599–2608. [CrossRef] [PubMed]
10. Estruch, R.; Ros, E.; Salas-Salvado, J.; Covas, M.I.; Corella, D.; Aros, F.; Gomez-Gracia, E.; Ruiz-Gutierrez, V.; Fiol, M.; Lapetra, J.; et al. Primary Prevention of Cardiovascular Disease with a Mediterranean Diet Supplemented with Extra-Virgin Olive Oil or Nuts. *N. Engl. J. Med.* **2018**, *378*, e34. [CrossRef]
11. Hernáez, Á.; Estruch, R. The Mediterranean Diet and Cancer: What Do Human and Molecular Studies Have to Say about It? *Nutrients* **2019**, *11*, 2155. [CrossRef] [PubMed]
12. Martín-Peláez, S.; Fito, M.; Castaner, O. Mediterranean Diet Effects on Type 2 Diabetes Prevention, Disease Progression, and Related Mechanisms. A Review. *Nutrients* **2020**, *12*, 2236. [CrossRef] [PubMed]
13. Franquesa, M.; Pujol-Busquets, G.; García-Fernández, E.; Rico, L.; Shamirian-Pulido, L.; Aguilar-Martínez, A.; Medina, F.X.; Serra-Majem, L.; Bach-Faig, A. Mediterranean Diet and Cardiodiabesity: A Systematic Review through Evidence-Based Answers to Key Clinical Questions. *Nutrients* **2019**, *11*, 655. [CrossRef]
14. Salas-Salvadó, J.; Bulló, M.; Estruch, R.; Ros, E.; Covas, M.I.; Ibarrola-Jurado, N.; Corella, D.; Arós, F.; Gómez-Gracia, E.; Ruiz-Gutiérrez, V.; et al. Prevention of diabetes with Mediterranean diets: A subgroup analysis of a randomized trial. *Ann. Intern. Med.* **2014**, *160*, 1–10. [CrossRef]
15. Klimova, B.; Novotny, M.; Schlegel, P.; Valis, M. The Effect of Mediterranean Diet on Cognitive Functions in the Elderly Population. *Nutrients* **2021**, *13*, 2067. [CrossRef] [PubMed]

16. Psaltopoulou, T.; Sergentanis, T.N.; Panagiotakos, D.B.; Sergentanis, I.N.; Kosti, R.; Scarmeas, N. Mediterranean diet, stroke, cognitive impairment, and depression: A meta-analysis. *Ann. Neurol.* **2013**, *74*, 580–591. [CrossRef]
17. Salvatore, F.P.; Relja, A.; Filipcic, I.S.; Polasek, O.; Kolcic, I. Mediterranean diet and mental distress: "10,001 Dalmatians" study. *Br. Food J.* **2019**, *121*, 1314–1326. [CrossRef]
18. Bonaccio, M.; Di Castelnuovo, A.; Bonanni, A.; Costanzo, S.; De Lucia, F.; Pounis, G.; Zito, F.; Donati, M.B.; de Gaetano, G.; Iacoviello, L. Adherence to a Mediterranean diet is associated with a better health-related quality of life: A possible role of high dietary antioxidant content. *BMJ Open* **2013**, *3*, e003003. [CrossRef] [PubMed]
19. Granado-Casas, M.; Martin, M.; Martínez-Alonso, M.; Alcubierre, N.; Hernández, M.; Alonso, N.; Castelblanco, E.; Mauricio, D. The Mediterranean Diet is Associated with an Improved Quality of Life in Adults with Type 1 Diabetes. *Nutrients* **2020**, *12*, 131. [CrossRef]
20. Jacka, F.N.; O'Neil, A.; Opie, R.; Itsiopoulos, C.; Cotton, S.; Mohebbi, M.; Castle, D.; Dash, S.; Mihalopoulos, C.; Chatterton, M.L.; et al. A randomised controlled trial of dietary improvement for adults with major depression (the 'SMILES' trial). *BMC Med.* **2017**, *15*, 23. [CrossRef]
21. Popkin, B.M.; Adair, L.S.; Ng, S.W. Global nutrition transition and the pandemic of obesity in developing countries. *Nutr. Rev.* **2012**, *70*, 3–21. [CrossRef]
22. Garcia-Closas, R.; Berenguer, A.; Gonzalez, C.A. Changes in food supply in Mediterranean countries from 1961 to 2001. *Public Health Nutr.* **2006**, *9*, 53–60. [CrossRef] [PubMed]
23. Tourlouki, E.; Matalas, A.L.; Bountziouka, V.; Tyrovolas, S.; Zeimbekis, A.; Gotsis, E.; Tsiligianni, I.; Protopapa, I.; Protopapas, C.; Metallinos, G.; et al. Are current dietary habits in Mediterranean islands a reflection of the past? Results from the MEDIS study. *Ecol. Food Nutr.* **2013**, *52*, 371–386. [CrossRef] [PubMed]
24. Tessier, S.; Gerber, M. Factors determining the nutrition transition in two Mediterranean islands: Sardinia and Malta. *Public Health Nutr.* **2005**, *8*, 1286–1292. [CrossRef]
25. Da Silva, R.; Bach-Faig, A.; Raido Quintana, B.; Buckland, G.; Vaz de Almeida, M.D.; Serra-Majem, L. Worldwide variation of adherence to the Mediterranean diet, in 1961–1965 and 2000–2003. *Public Health Nutr.* **2009**, *12*, 1676–1684. [CrossRef] [PubMed]
26. Alberti-Fidanza, A.; Fidanza, F. Mediterranean Adequacy Index of Italian diets. *Public Health Nutr.* **2004**, *7*, 937–941. [CrossRef]
27. Rodrigues, S.S.; Caraher, M.; Trichopoulou, A.; de Almeida, M.D. Portuguese households' diet quality (adherence to Mediterranean food pattern and compliance with WHO population dietary goals): Trends, regional disparities and socioeconomic determinants. *Eur. J. Clin. Nutr.* **2008**, *62*, 1263–1272. [CrossRef] [PubMed]
28. Tur, J.A.; Romaguera, D.; Pons, A. Food consumption patterns in a Mediterranean region: Does the Mediterranean diet still exist? *Ann. Nutr. Metab.* **2004**, *48*, 193–201. [CrossRef]
29. Alves, R.; Perelman, J. Dietary changes during the Great Recession in Portugal: Comparing the 2005/2006 and the 2014 health surveys. *Public Health Nutr.* **2019**, *22*, 1971–1978. [CrossRef]
30. Bibiloni Mdel, M.; Martinez, E.; Llull, R.; Pons, A.; Tur, J.A. Western and Mediterranean dietary patterns among Balearic Islands' adolescents: Socio-economic and lifestyle determinants. *Public Health Nutr.* **2012**, *15*, 683–692. [CrossRef] [PubMed]
31. Kolcic, I.; Relja, A.; Gelemanovic, A.; Miljkovic, A.; Boban, K.; Hayward, C.; Rudan, I.; Polasek, O. Mediterranean diet in the southern Croatia—Does it still exist? *Croat. Med. J.* **2016**, *57*, 415–424. [CrossRef]
32. Naja, F.; Hwalla, N.; Hachem, F.; Abbas, N.; Chokor, F.A.Z.; Kharroubi, S.; Chamieh, M.C.; Jomaa, L.; Nasreddine, L. Erosion of the Mediterranean diet among adolescents: Evidence from an Eastern Mediterranean Country. *Br. J. Nutr.* **2021**, *125*, 346–356. [CrossRef] [PubMed]
33. Veronese, N.; Notarnicola, M.; Cisternino, A.M.; Inguaggiato, R.; Guerra, V.; Reddavide, R.; Donghia, R.; Rotolo, O.; Zinzi, I.; Leandro, G.; et al. Trends in adherence to the Mediterranean diet in South Italy: A cross sectional study. *Nutr. Metab. Cardiovasc. Dis.* **2020**, *30*, 410–417. [CrossRef] [PubMed]
34. Peng, W.; Goldsmith, R.; Shimony, T.; Berry, E.M.; Sinai, T. Trends in the adherence to the Mediterranean diet in Israeli adolescents: Results from two national health and nutrition surveys, 2003 and 2016. *Eur. J. Nutr.* **2021**. ahead of print. [CrossRef]
35. Tong, T.Y.N.; Imamura, F.; Monsivais, P.; Brage, S.; Griffin, S.J.; Wareham, N.J.; Forouhi, N.G. Dietary cost associated with adherence to the Mediterranean diet, and its variation by socio-economic factors in the UK Fenland Study. *Br. J. Nutr.* **2018**, *119*, 685–694. [CrossRef] [PubMed]
36. Pastor, R.; Pinilla, N.; Tur, J.A. The Economic Cost of Diet and Its Association with Adherence to the Mediterranean Diet in a Cohort of Spanish Primary Schoolchildren. *Int. J. Environ. Res. Public Health* **2021**, *18*, 1282. [CrossRef]
37. Schröder, H.; Gomez, S.F.; Ribas-Barba, L.; Pérez-Rodrigo, C.; Bawaked, R.A.; Fíto, M.; Serra-Majem, L. Monetary Diet Cost, Diet Quality, and Parental Socioeconomic Status in Spanish Youth. *PLoS ONE* **2016**, *11*, e0161422. [CrossRef] [PubMed]
38. Lopez, C.N.; Martinez-Gonzalez, M.A.; Sanchez-Villegas, A.; Alonso, A.; Pimenta, A.M.; Bes-Rastrollo, M. Costs of Mediterranean and western dietary patterns in a Spanish cohort and their relationship with prospective weight change. *J. Epidemiol. Community Health* **2009**, *63*, 920–927. [CrossRef] [PubMed]
39. Bonaccio, M.; Bonanni, A.E.; Di Castelnuovo, A.; De Lucia, F.; Donati, M.B.; de Gaetano, G.; Iacoviello, L. Low income is associated with poor adherence to a Mediterranean diet and a higher prevalence of obesity: Cross-sectional results from the Moli-sani study. *BMJ Open* **2012**, *2*, e001685. [CrossRef] [PubMed]

40. Rippin, H.L.; Hutchinson, J.; Greenwood, D.C.; Jewell, J.; Breda, J.J.; Martin, A.; Rippin, D.M.; Schindler, K.; Rust, P.; Fagt, S.; et al. Inequalities in education and national income are associated with poorer diet: Pooled analysis of individual participant data across 12 European countries. *PLoS ONE* **2020**, *15*, e0232447. [CrossRef] [PubMed]
41. D'Innocenzo, S.; Biagi, C.; Lanari, M. Obesity and the Mediterranean Diet: A Review of Evidence of the Role and Sustainability of the Mediterranean Diet. *Nutrients* **2019**, *11*, 1306. [CrossRef] [PubMed]
42. Agnoli, C.; Sieri, S.; Ricceri, F.; Giraudo, M.T.; Masala, G.; Assedi, M.; Panico, S.; Mattiello, A.; Tumino, R.; Giurdanella, M.C.; et al. Adherence to a Mediterranean diet and long-term changes in weight and waist circumference in the EPIC-Italy cohort. *Nutr. Diabetes* **2018**, *8*, 22. [CrossRef] [PubMed]
43. Esposito, K.; Marfella, R.; Ciotola, M.; Di Palo, C.; Giugliano, F.; Giugliano, G.; D'Armiento, M.; D'Andrea, F.; Giugliano, D. Effect of a Mediterranean-style diet on endothelial dysfunction and markers of vascular inflammation in the metabolic syndrome: A randomized trial. *JAMA* **2004**, *292*, 1440–1446. [CrossRef] [PubMed]
44. Bouzas, C.; Bibiloni, M.D.M.; Jubilert, A.; Ruiz-Canela, M.; Salas-Salvadó, J.; Corella, D.; Zomeño, M.D.; Romaguera, D.; Vioque, J.; Alonso-Gómez, Á.M.; et al. Adherence to the Mediterranean Lifestyle and Desired Body Weight Loss in a Mediterranean Adult Population with Overweight: A PREDIMED-Plus Study. *Nutrients* **2020**, *12*, 2114. [CrossRef] [PubMed]
45. Koliaki, C.; Spinos, T.; Spinou, M.; Brinia, M.E.; Mitsopoulou, D.; Katsilambros, N. Defining the Optimal Dietary Approach for Safe, Effective and Sustainable Weight Loss in Overweight and Obese Adults. *Healthcare* **2018**, *6*, 73. [CrossRef]
46. Rehberg, J.; Stipcic, A.; Coric, T.; Kolcic, I.; Polasek, O. Mortality patterns in Southern Adriatic islands of Croatia: A registry-based study. *Croat. Med. J.* **2018**, *59*, 118–123. [CrossRef]
47. UNESCO. Mediterranean diet inscribed in 2013 on the Representative List of the Intangible Cultural Heritage of Humanity. Available online: http://www.unesco.org/culture/ich/en/RL/00884 (accessed on 10 September 2021).
48. World Health Organization. Noncommunicable Diseases (NCD) Country Profiles. 2018. Available online: https://www.who.int/nmh/countries/2018/hrv_en.pdf?ua=1 (accessed on 30 April 2021).
49. Gallus, S.; Lugo, A.; Murisic, B.; Bosetti, C.; Boffetta, P.; La Vecchia, C. Overweight and obesity in 16 European countries. *Eur. J. Nutr.* **2015**, *54*, 679–689. [CrossRef] [PubMed]
50. Musić Milanović, S.; Lang Morović, M.; Bukal, D.; Križan, H.; Buoncristiano, M.; Breda, J. Regional and sociodemographic determinants of the prevalence of overweight and obesity in children aged 7-9 years in Croatia. *Acta Clin. Croat.* **2020**, *59*, 303–311. [CrossRef]
51. Rudan, I.; Marusic, A.; Jankovic, S.; Rotim, K.; Boban, M.; Lauc, G.; Grkovic, I.; Dogas, Z.; Zemunik, T.; Vatavuk, Z.; et al. "10001 Dalmatians": Croatia launches its national biobank. *Croat. Med. J.* **2009**, *50*, 4–6. [CrossRef]
52. Relja, A.; Miljkovic, A.; Gelemanovic, A.; Boskovic, M.; Hayward, C.; Polasek, O.; Kolcic, I. Nut Consumption and Cardiovascular Risk Factors: A Cross-Sectional Study in a Mediterranean Population. *Nutrients* **2017**, *9*, 1296. [CrossRef]
53. Miljkovic, A.; Stipcic, A.; Bras, M.; Dordevic, V.; Brajkovic, L.; Hayward, C.; Pavic, A.; Kolcic, I.; Polasek, O. Is experimentally induced pain associated with socioeconomic status? Do poor people hurt more? *Med. Sci. Monit.* **2014**, *20*, 1232–1238. [CrossRef]
54. Monteagudo, C.; Mariscal-Arcas, M.; Rivas, A.; Lorenzo-Tovar, M.L.; Tur, J.A.; Olea-Serrano, F. Proposal of a Mediterranean Diet Serving Score. *PLoS ONE* **2015**, *10*, e0128594. [CrossRef] [PubMed]
55. Marendić, M.; Polić, N.; Matek, H.; Oršulić, L.; Polašek, O.; Kolčić, I. Mediterranean diet assessment challenges: Validation of the Croatian Version of the 14-item Mediterranean Diet Serving Score (MDSS) Questionnaire. *PLoS ONE* **2021**, *16*, e0247269. [CrossRef]
56. Leone, A.; Battezzati, A.; De Amicis, R.; De Carlo, G.; Bertoli, S. Trends of Adherence to the Mediterranean Dietary Pattern in Northern Italy from 2010 to 2016. *Nutrients* **2017**, *9*, 734. [CrossRef]
57. Bibiloni, M.D.M.; Gonzalez, M.; Julibert, A.; Llompart, I.; Pons, A.; Tur, J.A. Ten-Year Trends (1999-2010) of Adherence to the Mediterranean Diet among the Balearic Islands' Adult Population. *Nutrients* **2017**, *9*, 749. [CrossRef]
58. Quarta, S.; Massaro, M.; Chervenkov, M.; Ivanova, T.; Dimitrova, D.; Jorge, R.; Andrade, V.; Philippou, E.; Zisimou, C.; Maksimova, V.; et al. Persistent Moderate-to-Weak Mediterranean Diet Adherence and Low Scoring for Plant-Based Foods across Several Southern European Countries: Are We Overlooking the Mediterranean Diet Recommendations? *Nutrients* **2021**, *13*, 1432. [CrossRef] [PubMed]
59. Abellan Aleman, J.; Zafrilla Rentero, M.P.; Montoro-Garcia, S.; Mulero, J.; Perez Garrido, A.; Leal, M.; Guerrero, L.; Ramos, E.; Ruilope, L.M. Adherence to the "Mediterranean Diet" in Spain and Its Relationship with Cardiovascular Risk (DIMERICA Study). *Nutrients* **2016**, *8*, 680. [CrossRef]
60. Benhammou, S.; Heras-Gonzalez, L.; Ibanez-Peinado, D.; Barcelo, C.; Hamdan, M.; Rivas, A.; Mariscal-Arcas, M.; Olea-Serrano, F.; Monteagudo, C. Comparison of Mediterranean diet compliance between European and non-European populations in the Mediterranean basin. *Appetite* **2016**, *107*, 521–526. [CrossRef]
61. Dernini, S.; Berry, E.M.; Serra-Majem, L.; La Vecchia, C.; Capone, R.; Medina, F.X.; Aranceta-Bartrina, J.; Belahsen, R.; Burlingame, B.; Calabrese, G.; et al. Med Diet 4.0: The Mediterranean diet with four sustainable benefits. *Public Health Nutr.* **2017**, *20*, 1322–1330. [CrossRef] [PubMed]
62. Marino, M.; Puppo, F.; Del Bo, C.; Vinelli, V.; Riso, P.; Porrini, M.; Martini, D. A Systematic Review of Worldwide Consumption of Ultra-Processed Foods: Findings and Criticisms. *Nutrients* **2021**, *13*, 2778. [CrossRef] [PubMed]

63. Da Rocha, B.R.S.; Rico-Campa, A.; Romanos-Nanclares, A.; Ciriza, E.; Barbosa, K.B.F.; Martinez-Gonzalez, M.A.; Martin-Calvo, N. Adherence to Mediterranean diet is inversely associated with the consumption of ultra-processed foods among Spanish children: The SENDO project. *Public Health Nutr.* **2021**, *24*, 3294–3303. [CrossRef] [PubMed]
64. Affret, A.; Severi, G.; Dow, C.; Rey, G.; Delpierre, C.; Boutron-Ruault, M.C.; Clavel-Chapelon, F.; Fagherazzi, G. Socio-economic factors associated with a healthy diet: Results from the E3N study. *Public Health Nutr.* **2017**, *20*, 1574–1583. [CrossRef] [PubMed]
65. Kesse-Guyot, E.; Bertrais, S.; Peneau, S.; Estaquio, C.; Dauchet, L.; Vergnaud, A.C.; Czernichow, S.; Galan, P.; Hercberg, S.; Bellisle, F. Dietary patterns and their sociodemographic and behavioural correlates in French middle-aged adults from the SU.VI.MAX cohort. *Eur. J. Clin. Nutr.* **2009**, *63*, 521–528. [CrossRef]
66. Cavaliere, A.; De Marchi, E.; Banterle, A. Exploring the Adherence to the Mediterranean Diet and Its Relationship with Individual Lifestyle: The Role of Healthy Behaviors, Pro-Environmental Behaviors, Income, and Education. *Nutrients* **2018**, *10*, 141. [CrossRef]
67. Dinnissen, C.S.; Ocke, M.C.; Buurma-Rethans, E.J.M.; van Rossum, C.T.M. Dietary Changes among Adults in The Netherlands in the Period 2007-2010 and 2012-2016. Results from Two Cross-Sectional National Food Consumption Surveys. *Nutrients* **2021**, *13*, 1520. [CrossRef] [PubMed]
68. Darmon, N.; Drewnowski, A. Does social class predict diet quality? *Am. J. Clin. Nutr.* **2008**, *87*, 1107–1117. [CrossRef]
69. Darmon, N.; Drewnowski, A. Contribution of food prices and diet cost to socioeconomic disparities in diet quality and health: A systematic review and analysis. *Nutr. Rev.* **2015**, *73*, 643–660. [CrossRef]
70. Bonaccio, M.; Di Castelnuovo, A.; Pounis, G.; Costanzo, S.; Persichillo, M.; Cerletti, C.; Donati, M.B.; de Gaetano, G.; Iacoviello, L. High adherence to the Mediterranean diet is associated with cardiovascular protection in higher but not in lower socioeconomic groups: Prospective findings from the Moli-sani study. *Int. J. Epidemiol.* **2017**, *46*, 1478–1487. [CrossRef] [PubMed]
71. Teixeira, B.; Afonso, C.; Sousa, A.S.; Guerra, R.S.; Santos, A.; Borges, N.; Moreira, P.; Padrao, P.; Amaral, T.F. Adherence to a Mediterranean Dietary Pattern status and associated factors among Portuguese older adults: Results from the Nutrition UP 65 cross-sectional study. *Nutrition* **2019**, *65*, 91–96. [CrossRef]
72. Grosso, G.; Marventano, S.; Giorgianni, G.; Raciti, T.; Galvano, F.; Mistretta, A. Mediterranean diet adherence rates in Sicily, southern Italy. *Public Health Nutr.* **2014**, *17*, 2001–2009. [CrossRef]
73. Vitale, M.; Racca, E.; Izzo, A.; Giacco, A.; Parente, E.; Riccardi, G.; Giacco, R. Adherence to the traditional Mediterranean diet in a population of South of Italy: Factors involved and proposal of an educational field-based survey tool. *Int. J. Food Sci. Nutr.* **2019**, *70*, 195–201. [CrossRef] [PubMed]
74. Ferro-Luzzi, A.; James, W.P.; Kafatos, A. The high-fat Greek diet: A recipe for all? *Eur. J. Clin. Nutr.* **2002**, *56*, 796–809. [CrossRef] [PubMed]
75. Biesbroek, S.; Kneepkens, M.C.; van den Berg, S.W.; Fransen, H.P.; Beulens, J.W.; Peeters, P.H.M.; Boer, J.M.A. Dietary patterns within educational groups and their association with CHD and stroke in the European Prospective Investigation into Cancer and Nutrition-Netherlands cohort. *Br. J. Nutr.* **2018**, *119*, 949–956. [CrossRef]
76. Galobardes, B.; Morabia, A.; Bernstein, M.S. Diet and socioeconomic position: Does the use of different indicators matter? *Int. J. Epidemiol.* **2001**, *30*, 334–340. [CrossRef] [PubMed]
77. Cuschieri, S.; Libra, M. Adherence to the Mediterranean Diet in Maltese Adults. *Int. J. Environ. Res. Public Health* **2020**, *18*, 10. [CrossRef]
78. Novak, D.; Stefan, L.; Prosoli, R.; Emeljanovas, A.; Mieziene, B.; Milanovic, I.; Radisavljevic-Janic, S. Mediterranean Diet and Its Correlates among Adolescents in Non-Mediterranean European Countries: A Population-Based Study. *Nutrients* **2017**, *9*, 177. [CrossRef] [PubMed]
79. Tur, J.A.; Romaguera, D.; Pons, A. Adherence to the Mediterranean dietary pattern among the population of the Balearic Islands. *Br. J. Nutr.* **2004**, *92*, 341–346. [CrossRef] [PubMed]
80. Sanchez-Villegas, A.; Martinez, J.A.; De Irala, J.; Martinez-Gonzalez, M.A. Determinants of the adherence to an "a priori" defined Mediterranean dietary pattern. *Eur. J. Nutr.* **2002**, *41*, 249–257. [CrossRef]
81. Caparello, G.; Galluccio, A.; Giordano, C.; Lofaro, D.; Barone, I.; Morelli, C.; Sisci, D.; Catalano, S.; Ando, S.; Bonofiglio, D. Adherence to the Mediterranean diet pattern among university staff: A cross-sectional web-based epidemiological study in Southern Italy. *Int. J. Food. Sci. Nutr.* **2020**, *71*, 581–592. [CrossRef]
82. Grosso, G.; Marventano, S.; Buscemi, S.; Scuderi, A.; Matalone, M.; Platania, A.; Giorgianni, G.; Rametta, S.; Nolfo, F.; Galvano, F.; et al. Factors associated with adherence to the Mediterranean diet among adolescents living in Sicily, Southern Italy. *Nutrients* **2013**, *5*, 4908–4923. [CrossRef] [PubMed]
83. Mendez, M.A.; Popkin, B.M.; Jakszyn, P.; Berenguer, A.; Tormo, M.J.; Sanchez, M.J.; Quiros, J.R.; Pera, G.; Navarro, C.; Martinez, C.; et al. Adherence to a Mediterranean diet is associated with reduced 3-year incidence of obesity. *J. Nutr.* **2006**, *136*, 2934–2938. [CrossRef]
84. Shatwan, I.M.; Alhinai, E.A.; Alawadhi, B.; Surendran, S.; Aljefree, N.M.; Almoraie, N.M. High Adherence to the Mediterranean Diet Is Associated with a Reduced Risk of Obesity among Adults in Gulf Countries. *Nutrients* **2021**, *13*, 995. [CrossRef]
85. Romaguera, D.; Norat, T.; Mouw, T.; May, A.M.; Bamia, C.; Slimani, N.; Travier, N.; Besson, H.; Luan, J.; Wareham, N.; et al. Adherence to the Mediterranean diet is associated with lower abdominal adiposity in European men and women. *J. Nutr.* **2009**, *139*, 1728–1737. [CrossRef] [PubMed]

86. Cena, H.; Porri, D.; De Giuseppe, R.; Kalmpourtzidou, A.; Salvatore, F.P.; El Ghoch, M.; Itani, L.; Kreidieh, D.; Brytek-Matera, A.; Pocol, C.B.; et al. How Healthy Are Health-Related Behaviors in University Students: The HOLISTic Study. *Nutrients* **2021**, *13*, 675. [CrossRef]
87. Bonaccio, M.; Di Castelnuovo, A.; Bonanni, A.; Costanzo, S.; Persichillo, M.; Cerletti, C.; Donati, M.B.; de Gaetano, G.; Iacoviello, L. Socioeconomic status and impact of the economic crisis on dietary habits in Italy: Results from the INHES study. *J. Public Health* **2018**, *40*, 703–712. [CrossRef]
88. Bonaccio, M.; Di Castelnuovo, A.; Bonanni, A.; Costanzo, S.; De Lucia, F.; Persichillo, M.; Zito, F.; Donati, M.B.; de Gaetano, G.; Iacoviello, L. Decline of the Mediterranean diet at a time of economic crisis. Results from the Moli-sani study. *Nutr. Metab. Cardiovasc. Dis.* **2014**, *24*, 853–860. [CrossRef] [PubMed]
89. Kosti, R.I.; Kanellopoulou, A.; Notara, V.; Antonogeorgos, G.; Rojas-Gil, A.P.; Kornilaki, E.N.; Lagiou, A.; Panagiotakos, D.B. Household food spending, parental and childhood's diet quality, in financial crisis: A cross-sectional study in Greece. *Eur. J. Public Health* **2021**, *31*, 822–828. [CrossRef] [PubMed]
90. Bonaccio, M.; Bes-Rastrollo, M.; de Gaetano, G.; Iacoviello, L. Challenges to the Mediterranean diet at a time of economic crisis. *Nutr. Metab. Cardiovasc. Dis.* **2016**, *26*, 1057–1063. [CrossRef]
91. Burlingame, B.G.V.; Meybeck, A. *Mediterranean Food Consumption Patterns: Diet, Environment, Society, Economy and Health*; FAO, Ed.; CIHEAM/FAO: Rome, Italy, 2015.
92. Balanza, R.; Garcia-Lorda, P.; Perez-Rodrigo, C.; Aranceta, J.; Bonet, M.B.; Salas-Salvado, J. Trends in food availability determined by the Food and Agriculture Organization's food balance sheets in Mediterranean Europe in comparison with other European areas. *Public Health Nutr.* **2007**, *10*, 168–176. [CrossRef] [PubMed]
93. Blas, A.; Garrido, A.; Unver, O.; Willaarts, B. A comparison of the Mediterranean diet and current food consumption patterns in Spain from a nutritional and water perspective. *Sci. Total Environ.* **2019**, *664*, 1020–1029. [CrossRef]
94. Varela-Moreiras, G.; Avila, J.M.; Cuadrado, C.; del Pozo, S.; Ruiz, E.; Moreiras, O. Evaluation of food consumption and dietary patterns in Spain by the Food Consumption Survey: Updated information. *Eur. J. Clin. Nutr.* **2010**, *64* (Suppl. 3), S37–S43. [CrossRef]
95. Micha, R.; Khatibzadeh, S.; Shi, P.L.; Andrews, K.G.; Engell, R.E.; Mozaffarian, D.; Dis, G.B.D.N.C. Global, regional and national consumption of major food groups in 1990 and 2010: A systematic analysis including 266 country-specific nutrition surveys worldwide. *BMJ Open* **2015**, *5*. [CrossRef]
96. Fismen, A.S.; Smith, O.R.; Torsheim, T.; Rasmussen, M.; Pedersen Pagh, T.; Augustine, L.; Ojala, K.; Samdal, O. Trends in Food Habits and Their Relation to Socioeconomic Status among Nordic Adolescents 2001/2002–2009/2010. *PLoS ONE* **2016**, *11*, e0148541. [CrossRef]
97. Nasreddine, L.; Ayoub, J.J.; Hachem, F.; Tabbara, J.; Sibai, A.M.; Hwalla, N.; Naja, F. Differences in Dietary Intakes among Lebanese Adults over a Decade: Results from Two National Surveys 1997–2008/2009. *Nutrients* **2019**, *11*, 1738. [CrossRef] [PubMed]
98. Greene, M.W.; Roberts, A.P.; Fruge, A.D. Negative Association Between Mediterranean Diet Adherence and COVID-19 Cases and Related Deaths in Spain and 23 OECD Countries: An Ecological Study. *Front. Nutr.* **2021**, *8*, 591964. [CrossRef]
99. Kristic, I.; Pehlic, M.; Pavlovic, M.; Kolaric, B.; Kolcic, I.; Polasek, O. Coronavirus epidemic in Croatia: Case fatality decline during summer? *Croat. Med. J.* **2020**, *61*, 501–507. [CrossRef]
100. Roskam, A.J.; Kunst, A.E.; Van Oyen, H.; Demarest, S.; Klumbiene, J.; Regidor, E.; Helmert, U.; Jusot, F.; Dzurova, D.; Mackenbach, J.P. Comparative appraisal of educational inequalities in overweight and obesity among adults in 19 European countries. *Int. J. Epidemiol.* **2010**, *39*, 392–404. [CrossRef] [PubMed]
101. Talukdar, D.; Seenivasan, S.; Cameron, A.J.; Sacks, G. The association between national income and adult obesity prevalence: Empirical insights into temporal patterns and moderators of the association using 40 years of data across 147 countries. *PLoS ONE* **2020**, *15*, e0232236. [CrossRef] [PubMed]

 nutrients

MDPI

Article

Deprivation Index and Lifestyle: Baseline Cross-Sectional Analysis of the PREDIMED-Plus Catalonia Study

Josep Basora [1,2,3,*], Felipe Villalobos [4], Meritxell Pallejà-Millán [4], Nancy Babio [2,3,5,*], Albert Goday [3,6,7,8], María Dolores Zomeño [3,7,9], Xavier Pintó [3,10], Emilio Sacanella [3,11,12] and Jordi Salas-Salvadó [2,3,5]

1 Fundació Institut Universitari per a la Recerca a l'Atenció Primària de Salut Jordi Gol i Gurina (IDIAPJGol), 08007 Barcelona, Spain
2 Unitat de Nutrició Humana, Departament de Bioquímica i Biotecnologia, Universitat Rovira i Virgili, 43201 Reus, Spain; jordi.salas@urv.cat
3 Consorcio CIBER, M.P. Fisiopatología de la Obesidad y Nutrición (CIBERObn), Instituto de Salud Carlos III (ISCIII), 28029 Madrid, Spain; agoday@parcdesalutmar.cat (A.G.); mzomeno@imim.es (M.D.Z.); xpinto@bellvitgehospital.cat (X.P.); esacane@clinic.cat (E.S.)
4 Unitat de Suport a la Recerca Tarragona-Reus, Fundació Institut Universitari per a la Recerca a l'Atenció Primària de Salut Jordi Gol i Gurina (IDIAPJGol), 43202 Reus, Spain; fvillalobos@idiapjgol.info (F.V.); mpalleja@idiapjgol.info (M.P.-M.)
5 Institut d'Investigació Sanitària Pere i Virgili (IISPV), Hospital Universitari San Joan de Reus, 43204 Reus, Spain
6 Servei d'Endocrinologia, Hospital del Mar, 08003 Barcelona, Spain
7 Cardiovascular Risk and Nutrition Research Group, Hospital del Mar Medical Research Institute (IMIM), 08003 Barcelona, Spain
8 Departament de Medicina, Univerisitat Autònoma de Barcelona, 08193 Barcelona, Spain
9 Nutrició Humana i Dietètica, Facultat de Ciències de la Salut Blanquerna-Universitat Ramon Llull, 08022 Barcelona, Spain
10 Lipid Unit, Department of Internal Medicine, Bellvitge Biomedical Research Institute (IDIBELL)-Hospital Universitari de Bellvitge, 08908 L'Hospitalet de Llobregat, Spain
11 Department of Internal Medicine, Hospital Clínic, 08036 Barcelona, Spain
12 School of Medicine, University of Barcelona, 08036 Barcelona, Spain
* Correspondence: jbasora@idiapjgol.org (J.B.); nancy.babio@urv.cat (N.B.); Tel.: +34-934-824516 (J.B.); +34-977-759312 (N.B.)

Citation: Basora, J.; Villalobos, F.; Pallejà-Millán, M.; Babio, N.; Goday, A.; Zomeño, M.D.; Pintó, X.; Sacanella, E.; Salas-Salvadó, J. Deprivation Index and Lifestyle: Baseline Cross-Sectional Analysis of the PREDIMED-Plus Catalonia Study. *Nutrients* **2021**, *13*, 3408. https://doi.org/10.3390/nu13103408

Academic Editors: Daniela Bonofiglio and Jose Lara

Received: 10 August 2021
Accepted: 23 September 2021
Published: 27 September 2021

Abstract: This baseline cross-sectional analysis from data acquired in a sub-sample of the PREDIMED-Plus study participants aimed to evaluate the relation between the Composite Socioeconomic Index (CSI) and lifestyle (diet and physical activity). This study involved 1512 participants (759 (52.2%) women) between 55 and 80 years with overweight/obesity and metabolic syndrome assigned to 137 primary healthcare centers in Catalonia, Spain. CSI and lifestyle (diet and physical activity) were assessed. Multiple linear regression or multinomial regression were applied to the data. Cluster analysis was performed to identify dietary patterns. The multiple linear regression model showed that a high deprivation index was related to a higher consumption of refined cereals (11.98 g/d, p-value = 0.001) and potatoes (6.68 g/d, p-value = 0.001), and to a lower consumption of fruits (−17.52 g/d, p-value = 0.036), and coffee and tea (−8.03 g/d, p-value = 0.013). Two a posteriori dietary patterns were identified by cluster analysis and labeled as "healthy" and "unhealthy". In addition, the multinomial regression model showed that a high deprivation index was related to an unhealthy dietary pattern and low physical activity (OR 1.42 [95% CI 1.06–1.89]; p-value < 0.05). In conclusion, a high deprivation index was related to an unhealthy lifestyle (diet and physical activity) in PREDIMED-Plus study participants.

Keywords: deprivation index; lifestyle; diet; physical activity

1. Introduction

Lifestyles—diet and physical activity—impact on a wider range of health outcomes. A recent review concluded that more than 11 million deaths were attributable to dietary risk factors in 2017 [1]. In addition, physical activity is also a major and globally relevant determinant of health, likewise a lack of physical activity is considered a key risk factor for the development of chronic diseases and premature mortality [2].

Socioeconomic status (SES) is an important determinant of health. The relationship that exist between the socioeconomic status (SES) and several health outcomes is well known. Individuals with low SES are at higher risk of chronic diseases (physical and mental health diseases) and have lower life expectancy compared to those with high SES [3].

A relationship between SES and lifestyle have been largely documented. An individual can choose a healthy or an unhealthy lifestyle, but this choice is determined by their SES and other social determinants such as age, sex, or civil status, leading to poor or good health [4,5]. Therefore, lifestyle could mediate the relationship between SES and health [6].

Previous studies have reported an association between SES and other social determinants, and lifestyles. Regarding diet, for example, an association between a high level of education and higher consumption of fruits and vegetables has been reported [7]. Men consumed more dairy products, olives, nuts and seeds, red meat and processed food, sweets, eggs, alcohol and fast food compared to women, while women consumed more fruits; and men with low SES have a higher consumption of alcoholic beverages, compared to women [8]. On the other hand, studies that have evaluated the relationship between SES and other social determinants, and dietary patterns have observed inconsistent results [9–14]. Some studies have reported higher adherence to a healthy dietary pattern in older individuals with high SES [9–14]. On the contrary, studies have frequently reported an increased risk of adherence to an unhealthy dietary pattern in women, married and with family, non-workers, and with high education level [15–17]. In relation to physical activity, low levels have been observed with more prevalence in older individuals, men, married, and with low educational level and SES [18]. In addition, physical inactivity was observed to be more frequent in individuals living in neighborhoods with low SES and deprivation [19,20].

In studies of health inequalities, SES is most commonly operationalized as either education, social class or income, and often without providing a rationale for the choice of indicator. Education, social class, or income can have overlapping properties in relation to health [21]. The deprivation indices are instruments used to measure health inequalities at a population level. All of them are constructed based on different socioeconomic or demographic characteristics and are used to quantify the socioeconomic variation in health outcomes. They measure socioeconomic vulnerability and some of them make it possible to prioritize services and corrective actions. They started to be used in the 1980s in the United Kingdom (UK). The most well-known are the Townsend [22], and Carstairs and Morris indices [23], however, in the last few years other indices have been developed and validated such us the SIMD16 index in Scotland, the ID2007 in UK, the NZDEP2018 in New Zealand, the MEDEAS index in Spain [24], and the Composite Socioeconomic Index (CSI) in Catalonia, Spain [25].

The CSI is a deprivation index for the assignation of the budgets of the primary healthcare areas in Catalonia (Spain) valid both in urban and rural areas. The variables used to construct the index allow frequent updating and are representative at the territorial level of primary healthcare: exemption from pharmaceutical co-payment, income below €18,000/year, income higher than €100,000/year, manual occupations, low level of education, mortality before the age of 75 and potentially avoidable hospitalizations [25]. The components of the CSI have demonstrated an association between low socioeconomic level and high morbidity rates, high use of primary healthcare services, hospital and psychiatric care, as well as a greater use of drugs, especially for mental health problems [26]. However, there is no scientific evidence on the relationship between CSI with the lifestyle, dietary aspects nor with physical activity.

The deprivation index is considered a good instrument for classifying the SES. The CSI is built using different socioeconomic indicators, and those have shown overlapping properties when they are used individually to measure health inequalities. Demonstrating that the CSI is related to patterns of unhealthy lifestyles, diet and physical activity is important in order to use it as an instrument to design and prioritize lifestyle interventions at the community level, especially in primary healthcare areas. In addition, it would allow us to have a broad vision of how the socioeconomic contextual aspects of geographic location impact on health related to diet and physical activity. For example, neighborhood-level characteristics, such as the availability of healthy food, and the quality of the physical environment, have been proposed as determinants of the overweight and obesity prevalence [27].

Bearing in mind the aforementioned, this study aimed to evaluate the relation between CSI and lifestyle (diet and physical activity), in a sub-sample of the PREDIMED-Plus study participants.

2. Materials and Methods

2.1. Study Design

This study is a baseline cross-sectional analysis of data acquired in a sub-sample of participants enrolled in the PREDIMED-Plus study from Catalonia Health Centers. The PREDIMED-Plus study is an ongoing, multicenter, randomized, controlled clinical trial conducted in Spain involving participants between 55 and 80 years with overweight/obesity and metabolic syndrome for primary cardiovascular prevention.

The study protocol is detailed in http://predimedplus.com/, and the description of the cohort has been published elsewhere [28]. The protocol was written in accordance with the ethical principles and good clinical practices contained in the Declaration of Helsinki. This study was registered at the International Standard Randomized Controlled Trial (ISRCT; http://www.isrctn.com/ISRCTN89898870) with number 89898870. The respective Institutional Review Board of all study centers approved the study protocol and all participants provided written informed consent.

2.2. Participants

For the present study, we included baseline data of participants living in Catalonia (Spain) recruited and randomized from the following centers: (a) Institut Hospital del Mar d'Investigacions Mèdiques (IMIM) in Barcelona, (b) Hospital Sant Joan-IISPV/Atenció Primària in Reus, (c) Atenció Primària Metro Sur-Departament d'Aterioesclorosi de l'Hospital de Bellvitge in Barcelona, and (d) Hospital Clinic of Barcelona. The period of recruitment was from October 2013 to December 2016. Participants did not receive any type of compensation for participating in the study. The present analysis included 1512 participants (759 women) from 137 primary healthcare areas affiliated to these centers. Participants recording extreme total energy intakes (<500 or >3500 kcal/day in women or <800 or >4000 kcal/day in men) [29] and without information on CSI were excluded (n = 69).

2.3. Variables Determined

2.3.1. Socio-Demographic Variables

Participants self-reported socio-demographic data: age, sex, civil status, education level and employment status.

2.3.2. Composed Socioeconomic Index

The CSI was used to determine the deprivation index of the participants [25]. All primary healthcare areas registered in Catalonia (n = 398) have assigned a CSI. The CSI ranges from −0.01 to 5.68, and a higher value of the CSI implies higher deprivation index. For this study, we included the primary healthcare areas registered from the PREDIMED-plus study participants, the CSI ranges from −0.004 to 4.49. We classified the participants into two categories according to the CSI assigned to their corresponding registered primary healthcare area: high deprivation index (≥2.27 points) and low deprivation index (<2.27 points).

2.3.3. Anthropometric Measurements

Body weight, height and waist circumference (WC) were measured by trained staff and following the PREDIMED-Plus operations protocol. Weight and height were measured using calibrated scales with participants wearing light clothes and no shoes. BMI was calculated as body weight (kg) divided by height (meters) squared. WC was measured with anthropometric tape midway between the lowest rib and the iliac crest.

2.3.4. Dietary Intake and Adherence to the Energy-Reduced Mediterranean Diet

A trained dietician asked the participants about their frequency of consumption for a specified serving size of each 143 items food frequency questionnaire item during the preceding year in a face-to-face interview [30]. For each item, a typical portion size was included, and consumption frequencies were registered in 9 categories: never or almost never, 1–3 times/month, once per week, 2–4 times/week, 5–6 times/week, once per day, 2–3 times/day, 4–6 times/day, and >6 times/day. Reported frequencies of food consumption were converted into frequencies per day, and multiplied by the weight of the typical portion size indicated to obtain the intake in g/d. To identify dietary patterns, 143 food items from the questionnaire were categorized in 23 food groups (Supplementary Table S1).

Adherence to the energy-reduced Mediterranean Diet (er-MedDiet) was assessed by trained dieticians using a recently validated questionnaire of 17 items [31]. This questionnaire has been used in the ongoing PREDIMED-Plus study aiming to assess the effect of an er-MedDiet on cardiovascular events in people with overweight and obesity at increased risk of CVD. The er-MedDiet questionnaire includes 14 items on food consumption and three items on eating behaviors, with some of the items belonging to the MEDAS validated questionnaire measuring adherence to Mediterranean diet in the PREDIMED study [32].

2.3.5. Physical Activity

Physical activity was measured using the Minnesota Questionnaire validated for the Spanish population [33,34]. Intensity (light, moderate, or vigorous), frequency (days per week) and duration of physical activity (minutes per day) were registered. The intensity and frequency of each activity was used to calculate the intensity category in terms of metabolic equivalents (METs)/min/week. These values were obtained by multiplying the average energy expenditure (3.3 MET for walking, 4.0 MET for moderate intensity, and 8.0 MET for vigorous intensity) by min/week for each physical activity category. The results of each category of activity intensity were summed to obtain the total physical activity. Based on total physical activity, participants were classified into two categories: low physical activity ($\leq$2100 METs/min/week) and high physical activity (>2100 METs/min/week).

2.3.6. Sedentary Lifestyle, Smoking Habits and Clinical Morbidities

Sedentary lifestyle was measured using the Nurses' Health Study questionnaire validated for the Spanish population [35], consisting of a set of questions assessing the average daily time spent over the last year watching TV, sitting while using the computer, sitting during journeys, and total sitting. Answers included 12 categories ranging from never to $\geq$9 h/day of sitting time for the corresponding activity. A sedentary lifestyle was defined as $\geq$7 h/day of sitting time. Furthermore, participants reported their average daily sleeping time for both weekdays and weekends, using the non-validated open question, "How many hours do you sleep on average per day on weekdays and weekends?" Additional information related to smoking habits and clinical morbidities (presence of self-reported hypertension, dyslipidemia and type 2 diabetes mellitus) was collected.

2.4. Statistical Analysis

The data are presented as mean and standard deviation (SD) for continuous variables, or as a median and interquartile range [IR] for non-normally distributed data, and frequencies and percentages for categorical variables. Variables of the study were compared across different groups: CSI, food groups and lifestyles categories. We used *t*-tests or ANOVA-tests for comparisons of continuous variables among groups. The Mann–Whitney U test or the Kruskall–Wallis test was employed for the continuous variables that did not have a normal distribution according to the Kolmogorov–Smirnov test. For the pairwise comparison, corrected for multiple comparisons, the Tukey method was used when explanatory variables were normal-distributed and the Benjamini and Hochberg method otherwise. Comparisons among groups for categorical variables were performed with the χ^2 test and Fisher test when the expected frequencies were less than five.

The calorie-adjusted nutrient intake was made to avoid bias produced by the inter-individuals' variability of energy intake [36].

The relation between the CSI categories (low/high deprivation index) as exposure and the food consumption (food group, g/day) as outcome, was evaluated by multiple linear regression models adjusted by age (years), sex (man/woman), smoking (smoker, former smoker or never smoked), waist circumference (cm), physical activity (low/high), sedentary lifestyle (no/yes), hypertension (no/yes), dyslipidemia (no/yes) and type 2 diabetes mellitus (no/yes).

Cluster analysis, using the K-means method was performed to derive dietary patterns. The K-means method was applied based on Euclidean distances, and the data was input as z-scores. Two clusters were specified prior to analysis. Participants were divided based on the similarity of their food consumption (food groups adjusted for standardized energy).

By combining the dietary patterns created by cluster analysis ("healthy" and "unhealthy"), and the physical activity categories (low/high), four categories were created reflecting the lifestyle of the participants. In this way, each group of participants had to accomplish both conditions: be in the specified dietary pattern and specified physical activity category.

The relation between the CSI categories (low/high deprivation index) as outcome and the lifestyle (identified dietary pattern and physical activity category) as exposure, was evaluated by a multinomial regression model adjusted by age (years), sex (man/woman), smoking (smoker, former smoker or never smoked), waist circumference (cm), sedentary lifestyle (no/yes), hypertension (no/yes), dyslipidemia (no/yes), and type 2 diabetes mellitus (no/yes).

The selection of the covariates, which were included in the models, was based on the factors affecting choice of a healthy lifestyle [37], and on the inclusion criteria of the study, comorbidities could previously condition the sample for having received lifestyle interventions based on their risk factor.

Statistical significance was set at p-value < 0.05. Analyses were performed with the statistical software "R 4.03" for Windows.

3. Results

3.1. General Characteristics of the Participants and the Composite Socioeconomic Index (CSI) Categories

Table 1 shows the general characteristics of the participants according to the CSI categories. There were significant differences with respect to age, education level, employment status, adherence to the erMedDiet, physical activity, sedentary lifestyle and hypertension. Specifically, a higher percentage of participants with a high deprivation index compared to those with a low deprivation index, had a lower educational level and were not currently working. In addition, they had lower adherence to the erMedDiet, practiced less light and total physical activity, and a higher percentage of them had a sedentary lifestyle and hypertension.

Table 1. General characteristics of participants according to composed socioeconomic index categories.

	All	High Deprivation Index (≥2.27 Points)	Low Deprivation Index (<2.27 Points)	*p*-Value [#]
	n = 1512	*n* = 744	*n* = 768	
Socio-demographic variables				
Women	759 (50.2%)	378 (50.8%)	381 (49.6%)	0.679
Age (years)	65.5 (4.80)	65.3 (4.83)	65.8 (4.77)	0.041
Civil status *				
Single or religious	65 (4.31%)	32 (4.31%)	33 (4.31%)	0.794
Married	1145 (75.9%)	569 (76.6%)	576 (75.2%)	
Divorced or widowed	299 (19.8%)	142 (19.1%)	157 (20.5%)	
Education level *				
Academic or graduate	345 (22.9%)	113 (15.3%)	232 (30.2%)	<0.001
Secondary education	480 (31.9%)	213 (28.9%)	267 (34.8%)	
Primary education or less	680 (45.2%)	412 (55.8%)	268 (34.9%)	
Employment status *				
Currently working	304 (20.2%)	143 (19.3%)	161 (21.1%)	0.002
Disability	20 (1.33%)	14 (1.89%)	6 (0.79%)	
Housework	147 (9.77%)	92 (12.4%)	55 (7.20%)	
Retired	955 (63.5%)	450 (60.8%)	505 (66.1%)	
Unemployed	78 (5.19%)	41 (5.54%)	37 (4.84%)	
Anthropometric measurements				
BMI *				
Mean Kg/m^2	32.4 [30.1;35.1]	32.4 [30.2;34.9]	32.4 [30.0;35.4]	0.883
≥27 Kg/m^2	1494 (99.6%)	729 (99.6%)	765 (99.6%)	1.000
Waist circumference *				
Men (cm)	110 [104;116]	110 [104;116]	111 [105;118]	0.136
Women (cm)	104 [98.1;111]	104 [98.0;110]	105 [98.5;112]	0.084
Central obesity	1404 (93.5%)	689 (93.7%)	715 (93.2%)	0.761
Lifestyle				
Adherence to the erMedDiet (score from 0 to 17 points)	7.86 (2.51)	7.99 (2.52)	7.74 (2.50)	0.046
Physical activity (METs/Min/week)				
Light	839 [224;1678]	671 [112;1343]	839 [280;1678]	<0.001
Moderate	140 [0.00;1119]	140 [0.00;1171]	43.7 [0.00;1084]	0.196
Vigorous	83.9 [0.00;1119]	72.3 [0.00;934]	112 [0.00;1259]	0.695
Total	2098 [1105;3525]	1979 [1069;3357]	2241 [1133;3776]	0.044
Low physical activity	763 (50.5%)	394 (53.0%)	369 (48.0%)	0.063
High physical activity	749 (49.5%)	350 (47.0%)	399 (52.0%)	
Sedentary lifestyle	675 (44.7%)	299 (40.3%)	376 (49.9%)	0.001
Daily sleeping time (h/day) *	7.00 [6.00;8.00]	7.00 [6.00;8.00]	7.00 [6.00;8.00]	0.583
Smoking *				0.972
Current smoker	171 (11.3%)	84 (11.3%)	87 (11.4%)	
Former smoker	617 (40.9%)	302 (40.6%)	315 (41.1%)	
Never smoked	722 (47.8%)	358 (48.1%)	364 (47.5%)	
Comorbidities				
Dyslipidemia *	1042 (69.0%)	520 (70.0%)	522 (68.0%)	0.428
Hypertension	1318 (87.2%)	635 (85.3%)	683 (88.9%)	0.045
Type 2 diabetes mellitus	438 (29.0%)	204 (27.4%)	234 (30.5%)	0.211

Abbreviations: BMI: Body index mass; erMedDiet: energy reduced Mediterranean diet; METs: Metabolic Equivalents. Central obesity: waist circumference men >102 cm and women >88 cm. Data are presented as mean (SD) or median [IR] for continuous variables, and as *n* (%) for categorical variables. * The percentage of missing values was between 0.13% and 0.79% from total study population. *p*-value [#]: *t*-tests or Mann–Whitney U test for continuous variables; and χ^2 test and Fisher test for categorical variables.

3.2. Food Consumption of the Participants and the CSI Categories

Table 2 shows the food consumption of the participants according to the CSI categories. Participants with a high deprivation index compared to those with a low deprivation index had significantly lower consumption of full-fat dairy, red meat and meat products, whole grain cereals, fruits, sugar-free beverages, coffee or tea, spirits beverages and wines, and significantly higher consumption of refined cereals, potatoes, and biscuits and pastries.

Table 2. Food consumption of participants according to Composite Socioeconomic Index (CSI) categories.

	All	High Deprivation Index (≥2.27 Points)	Low Deprivation Index (<2.27 Points)	p-Value [#]
	n = 1512	n = 744	n = 768	
Full-fat dairy (g/day)	62.1 [34.6;115]	58.0 [32.6;111]	66.4 [37.0;121]	0.021
Low-fat dairy (g/day)	208 [83.0;319]	204 [68.1;315]	213 [98.3;322]	0.082
White meat (g/day)	68.5 [40.9;84.6]	70.2 [42.2;84.7]	67.6 [39.6;84.2]	0.686
Red meat and meat products (g/day)	88.0 [65.6;120]	85.9 [63.9;115]	92.0 [67.6;123]	0.022
Eggs (g/day)	25.1 [22.0;26.7]	25.0 [21.6;26.6]	25.2 [22.4;26.8]	0.065
Fish and seafood (g/day)	109 [74.8;143]	108 [75.5;143]	109 [73.7;143]	0.963
Whole grain cereals (g/day)	7.39 [0.22;73.9]	5.22 [0.00;67.7]	10.8 [0.38;74.1]	0.046
Refined cereals (g/day)	103 [61.1;155]	108 [63.4;167]	98.3 [59.5;143]	0.004
Legumes (g/day)	17.4 [13.0;24.1]	17.9 [13.3;24.2]	17.0 [12.9;24.1]	0.100
Fruits (g/day)	299 [200;401]	289 [194;377]	311 [209;413]	0.008
Vegetables (g/day)	305 [233;395]	301 [234;396]	310 [231;394]	0.918
Nuts (g/day)	8.87 [3.90;20.4]	9.00 [4.32;20.0]	8.75 [3.58;20.6]	0.378
Olive oil and olives (g/day)	56.5 [45.1;66.8]	55.1 [44.4;66.8]	57.2 [45.9;66.7]	0.175
Other fat or oils, full-fat dairy derivatives and processed meal (g/day)	6.46 [3.13;11.7]	6.49 [3.15;11.9]	6.36 [3.11;11.4]	0.895
Potatoes (g/day)	90.7 [47.2;103]	92.0 [51.8;104]	90.3 [42.9;102]	0.014
Biscuits and pastries (g/day)	14.4 [6.33;27.0]	15.1 [6.77;29.1]	13.8 [6.02;24.3]	0.023
Sugar, sweets, chocolate and cocoa (g/day)	14.4 [6.13;26.9]	14.8 [6.12;27.3]	14.2 [6.14;26.7]	0.831
Sugary beverages and juices (g/day)	35.4 [9.19;112]	36.7 [8.93;104]	34.6 [9.67;115]	0.750
Sugar-free beverages (g/day)	1.29 [0.00;13.3]	1.03 [0.00;11.6]	1.67 [0.00;15.0]	0.002
Coffee and tea (g/day)	92.9 [48.7;127]	71.0 [47.3;126]	100 [50.4;129]	0.005
Cava and beers (g/day)	50.7 [11.9;119]	48.4 [7.45;117]	53.5 [16.5;121]	0.129
Spirits beverages (g/day)	1.05 [0.00;3.10]	0.91 [0.00;3.11]	1.26 [0.00;3.09]	0.040
Wines (g/day)	27.6 [5.65;76.0]	24.4 [1.98;69.5]	30.6 [8.19;78.9]	0.014

Data are presented as median [IR]. p-value [#]: Mann–Whitney U test.

Multiple linear regression models showed that being a participant with a high deprivation index was related to a higher consumption of refined cereals (p-value = 0.001) and potatoes (p-value = 0.001), and to a lower consumption of fruits (p-value = 0.035), and coffee and tea (p-value = 0.012). No significant relationships were observed between the CSI categories and the consumption of other predefined food groups (Figure 1 and Supplementary Table S2).

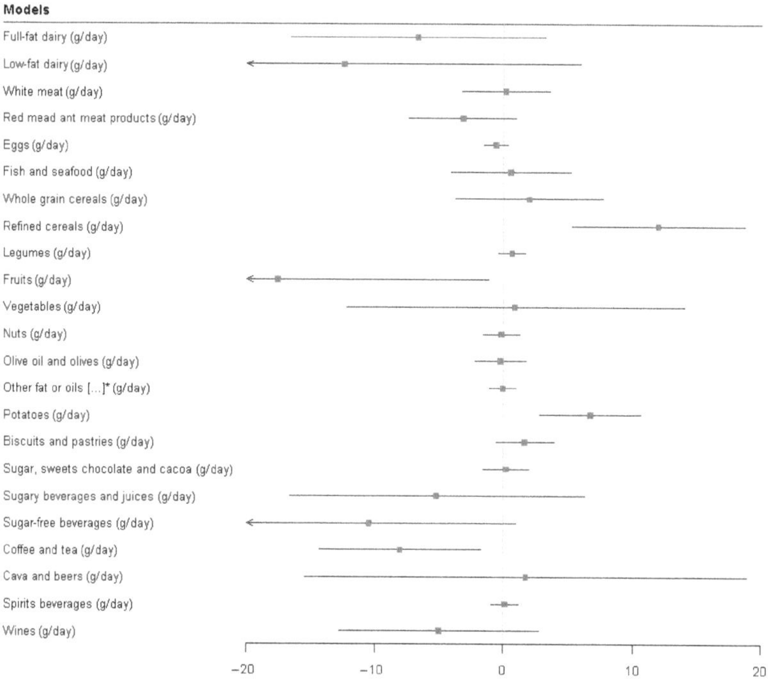

Figure 1. Relationship between CSI and food groups consumption. * other fat or oils, full-fat dairy derivatives and processed meal Multiple linear regressions. CSI (low high deprivation index) as exposure and food consumption (food groups, g/d) as outcome adjusted by age (years), sex (men/women). Somking (smoker, former or never smoked), waist circumference (cm). physical activity (low/high), sedentary lifestyle (no/yes), hypertension(no/yes), dyslipidemia (no/yes), dyslipidemia (no/yes), and type 2 diabetes mellitus.

3.3. Dietary Patterns

Figure 2 shows the two identified dietary patterns by cluster analysis. Due to their characteristics and affinities, these two patterns have been labeled as "healthy" followed by 704 (46.5%) participants, and "unhealthy" followed by 808 (53.5%). The "healthy" pattern was characterized by a significantly higher consumption of low-fat dairy, white meat, fish and seafood, whole grain cereals, legumes, fruits, vegetables, nuts, olive oil and olives. The "unhealthy" pattern was characterized by a significantly higher consumption of foods rich in fat, sugar and alcohol such as full-fat dairy, red meat and meat products, refined cereals, other fat or oils different from olive oil, full-fat dairy derivatives and processed meals, potatoes, biscuits and pastries, sugar, sweets, chocolate and cocoa, cava and beers, spirits and wines (Supplementary Table S3).

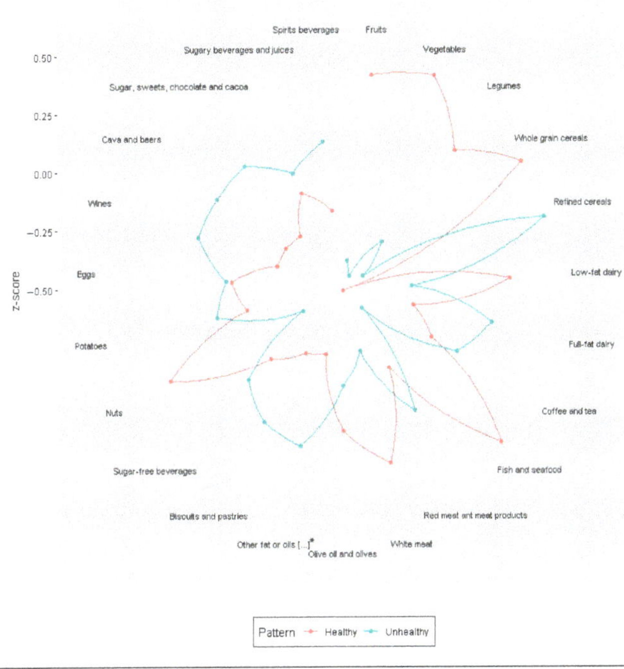

*Other fat or oils, full-fat dairy derivatives and processed meal.

Figure 2. Dietary patterns identified by cluster analysis.

3.4. General Characteristics of the Participants According to Predefined Lifestyle Categories

Four categories were created reflecting the lifestyle of the participants: (1) unhealthy dietary pattern and low physical activity, (2) unhealthy dietary pattern and high physical activity, (3) healthy dietary pattern and low physical activity, and (4) healthy dietary pattern and high physical activity.

Table 3 shows the general characteristics of the participants according to the predefined lifestyle categories. Significant differences were observed with respect to sex, age, employment status, BMI, waist circumference, central obesity, adherence to the erMedDiet, sedentary lifestyle, and smoking. Specifically, compared to those participants with an unhealthy dietary pattern & low physical activity, participants with a healthy dietary pattern & high physical activity were older, and were more likely to be women and retired, had lower BMI and waist circumference (men), and a lower percentage of them had central obesity; in addition, they had a higher adherence to the erMedDiet, and a lower percentage of them were sedentary and smokers.

Table 3. General characteristics of the participants according to the lifestyle categories.

	Unhealthy Dietary Pattern and Low Physical Activity	Unhealthy Dietary Pattern and High Physical Activity	Healthy Dietary Pattern and Low Physical Activity	Healthy Dietary Pattern and High Physical Activity	*p*-Overall [#]
	n = 435	*n* = 373	*n* = 328	*n* = 376	
Socio-demographic variables					
Women	192 (44.1%)	118 (31.6%) [a]	224 (68.3%) [a]	225 (59.8%) [a,b,c]	<0.001
Age (years)	65.1 (5.08)	65.1 (5.05)	65.8 (4.63)	66.1 (4.27) [a,b]	0.003
Civil status		a	b		0.024
Single or religious	20 (4.61%)	13 (3.49%)	15 (4.57%)	17 (4.55%)	
Married	312 (71.9%)	307 (82.3%)	239 (72.9%)	287 (76.7%)	
Divorced or widowed	102 (23.5%)	53 (14.2%)	74 (22.6%)	70 (18.7%)	
Education level *					0.112
Academic or graduate	108 (24.8%)	90 (24.2%)	68 (20.8%)	79 (21.3%)	
Secondary education	149 (34.3%)	124 (33.3%)	91 (27.8%)	116 (31.3%)	
Illiterate or primary education	178 (40.9%)	158 (42.5%)	168 (51.4%)	176 (47.4%)	
Employment status *		a	a	a,b,c	0.001
Currently working	129 (29.8%)	64 (17.3%)	62 (19.0%)	49 (13.1%)	
Disability	7 (1.62%)	4 (1.08%)	9 (2.76%)	0 (0.00%)	
Housework	39 (9.01%)	27 (7.30%)	36 (11.0%)	45 (12.0%)	
Retires	241 (55.7%)	249 (67.3%)	200 (61.3%)	265 (70.7%)	
Unemployed	17 (3.93%)	26 (7.03%)	19 (5.83%)	16 (4.27%)	
CSI					0.270
High deprivation index OYWX(≥2.27 points)	227 (52.2%)	177 (47.5%)	167 (50.9%)	173 (46.0%)	
Low deprivation index OYWX(<2.27 points)	208 (47.8%)	196 (52.5%)	161 (49.1%)	203 (54.0%)	
CSI (score)	2.48 [1.83;3.26]	2.26 [1.83;3.22]	2.40 [1.94;3.07]	2.26 [1.78;3.12]	0.436
Anthropometric measurements					
BMI *					
Kg/m^2	33.1 [30.3;35.8]	31.6 [29.6;34.2] [a]	33.0 [30.5;36.1] [b]	32.1 [30.0;34.5] [a,c]	<0.001
≥ 27 Kg/m^2	432 (100%)	371 (99.7%)	321 (99.1%)	370 (99.5%)	0.169
Waist circumference *					
Men (cm)	113 [106;119]	109 [104;115] [a]	111 [106;116]	109 [104;116] [a]	<0.001
Women (cm)	106 [98.2;114]	104 [97.6;110]	106 [100;111]	103 [97.0;109]	0.053
Central obesity	411 (94.7%)	331 (88.7%) [a]	311 (96.3%) [b]	351 (94.4%) [a,b]	<0.001
Lifestyle					
Adherence to erMedDiet (score)	7.00 [5.00;8.00]	7.00 [5.00;8.00]	9.00 [7.00;10.0] [a,b]	9.00 [8.00;11.0] [a,b]	<0.001
Sedentary lifestyle	226 (52.0%)	157 (42.1%) [a]	159 (48.8%) [b]	133 (35.4%) [a,b]	<0.001
Daily sleeping time (h/day) *	7.00 [6.00;8.00]	7.00 [6.00;8.00]	7.00 [6.00;8.00]	7.00 [6.00;8.00]	0.743
Smoking *		a	a,b	b	<0.001
Smoker	65 (15.0%)	48 (12.9%)	28 (8.54%)	30 (8.00%)	
Former smoker	172 (39.6%)	182 (48.8%)	108 (32.9%)	155 (41.3%)	
Never smoked	197 (45.4%)	143 (38.3%)	192 (58.5%)	190 (50.7%)	
Clinical morbidities					
Dyslipidemia *	285 (65.5%)	248 (66.5%)	235 (71.9%)	274 (72.9%)	0.059
Hypertension	386 (88.7%)	334 (89.5%)	273 (83.2%)	325 (86.4%)	0.056
Type 2 diabetes mellitus	114 (26.2%)	106 (28.4%)	111 (33.8%)	107 (28.5%)	0.139

BMI: Body index mass; erMedDiet: energy reduced Mediterranean diet; METs: Metabolic Equivalents. Central obesity: waist circumference men >102 cm and women >88 cm. Data are presented as mean (SD) or median [IR] for continuous variables, and as *n* (%) for categorical variables. * The percentage of missing values was between 0.13% and 0.79% from total study population. *p*-overall [#]: ANOVA or Kruskall–Wallis test for continuous variables, and χ^2 test or Fisher for categorical variables. For post-hoc comparisons: Tukey or Benjamini and Hochberg. [a], significant differences (*p*-value < 0.05) between-groups: *ref.* unhealthy pattern and low physical activity category. [b], significant differences (*p*-value < 0.05) between-groups: *ref.* unhealthy dietary pattern and high physical activity. [c], significant differences (*p*-value < 0.05) between-groups: *ref.* healthy pattern and low physical activity category.

3.5. Relation between CSI and Lifestyle

Table 4 shows the relationships between the CSI categories and the participant's lifestyle. The multinomial regression model shows that being a participant with a high deprivation index was positively related to a lifestyle composed of an unhealthy dietary pattern and low physical activity (OR 1.42 [95% CI 1.06–1.89]; p-value < 0.05). No significant associations were observed between the CSI categories and other predefined lifestyles considered.

Table 4. Relationship between CSI categories and lifestyle (dietary patterns + physical activity).

	Healthy Dietary Pattern and High Physical Activity	Unhealthy Dietary Pattern and Low Physical Activity OR [95% CI]	Unhealthy Dietary Pattern and High Physical Activity OR [95% CI]	Healthy Dietary Pattern and Low Physical Activity OR [95% CI]
CSI (high deprivation index)	ref.	1.42 [1.06,1.89] *	1.09 [0.81,1.48]	1.24 [0.91,1.68]
Sex (women)	ref.	0.66 [0.47,0.94] *	0.31 [0.22,0.46] **	1.60 [1.10,2.34] *
Age (years)	ref.	0.97 [0.94,1.00]	0.97 [0.94,1.00]	0.98 [0.94,1.01]
Smoking (former smoker)	ref.	0.47 [0.28,0.79] *	0.67 [0.40,1.14]	0.72 [0.40,1.29]
Smoking (never smoked)	ref.	0.67 [0.39,1.12]	0.79 [0.46,1.37]	0.97 [0.54,1.76]
Waist circumference (cm)	ref.	1.03 [1.03,1.01] **	1.00 [0.98,1.02]	1.01 [1.00,1.03]
Sedentary lifestyle (yes)	ref.	1.81 [1.34,2.43] **	1.20 [0.88,1.63]	1.75 [1.28,2.39]
Hypertension (yes)	ref.	1.33 [0.85,2.06]	1.39 [0.88,2.20]	0.82 [0.53,1.26]
Dyslipidemia (yes)	ref.	0.79 [0.57,1.08]	0.87 [0.63,1.20]	0.93 [0.66,1.31]
Type 2 diabetes mellitus (yes)	ref.	0.77 [0.56,1.07]	0.89 [0.63,1.24]	1.21 [0.86,1.69]

Multinomial regression model. CSI (low/high deprivation index) as exposure and lifestyle (healthy dietary pattern and high physical activity, unhealthy dietary pattern and low physical activity, unhealthy dietary pattern and high physical activity, healthy dietary pattern and low physical activity) as outcome, adjusted by age (years), sex (men/women), smoking (smoker, former or never smoked), waist circumference (cm), sedentary lifestyle (no/yes), hypertension (no/yes), dyslipidemia (no/yes) y type 2 diabetes mellitus (no/yes). Odds ratio (OR) and 95% confidence interval [CI] are shown. ** $p < 0.001$; * p-value < 0.05.

4. Discussion

In this baseline cross-sectional study conducted in PREDIMED-Plus study participants living in Catalonia, being a participant with a high deprivation index was related to a high consumption of refined cereals, potatoes, and to a lower consumption of fruits, and coffee and tea; four lifestyle categories were identified, and a high deprivation index was related with an "unhealthy" dietary pattern associated with low physical activity. These results support the limited existing evidence on the relationship between the deprivation index of a certain population area, dietary consumption and physical activity in individuals with overweight/obesity and metabolic syndrome.

Previous studies have observed that a high deprivation index is related to food consumption. A systematic review reported that individuals living in the most deprived areas had a lower consumption of fruits and vegetables [38]. Findings on relationship between measures of deprivation index and diet patterns have been inconsistent. In Australia, one study did not find any relationship between the deprivation index and the observed dietary patterns: Mediterranean, Prudent, and Western [39]; by contrast, in another study adherence to a healthy pattern (characterized by breakfast cereal, low fat milk, soy and rice milk, soup and stock, yoghurt, bananas, apples, other fruit and tea, and low consumption of pastries, potato chips, white bread, take-away foods, soft drinks, beer and wine) was inversely related to the deprivation index [10]. In a study conducted in Japan, individuals who lived in the most deprived areas had a lower score of adherence to the Japanese diet (low consumption of grains, potatoes, vegetables, fruits, mushrooms, fish, seafood; and high consumption of legumes, meat and coffee) [40].

Furthermore, our results were consistent with studies that have reported a relationship between the deprivation index and lifestyle. A study carried out in the UK observed that individuals in the highest deprivation quintile had a greater prevalence risk of adhering to an unhealthy lifestyle (low consumption of olive oil, fish, fruits and vegetables, high consumption of red and processed meat, and low physical activity) [41]. Similar observations were reported in a study in Australia, where a high deprivation index was related to an unhealthy lifestyle (less than five rations/day of fruits and vegetables, high alcohol consumption and low physical activity) [42].

The possible mechanisms related to being an individual with a high deprivation index and an "unhealthy" dietary pattern could be that individuals living in the most deprived areas suffer from so-called "food deserts" [43]. These areas are characterized by poor access to healthy and affordable food, and are characterized by social and spatial disparities in diet and diet-related health outcomes such as obesity [43]. A systematic review reported that better food access (availability, accessibility, affordability, accommodation and acceptability) is related to a healthy diet [44]. With respect to physical activity, it is recognized that green spaces accessibility may influence physical activity adherence [45]. The accessibility of greens spaces is usually better in more deprived areas but those residents have more negative perceptions (poorer perceived accessibility and poorer safety) and are less likely to use the green spaces [46].

One aspect that we can highlight from our findings is that being a woman is related with a healthy lifestyle. Previous studies support this relationship: women have reported higher adherence to a healthier diet [9–14], and higher levels of physical activity [18]. Women place greater importance on healthy eating than men, health beliefs explain a large proportion of dietary behavior, and they are more interested in and actively seek health-related information to a larger extent than men [47,48]. However other studies have reported an increased risk of adherence to an unhealthy dietary pattern in women [15–17]. More knowledge on gender differences in lifestyles could facilitate public health initiatives to promote healthy lifestyles in women and men.

A major strength of this study is the use of the CSI. It is a deprivation index that is valid for all basic health areas (urban and rural), it is easy to interpret, can be updated more frequently than indices constructed from census variables, and is related to the need of health service use, so it can be utilized to design and prioritize lifestyle interventions in primary healthcare, with community repercussions.

Some limitations deserve to be mentioned, such as the inherent nature of cross-sectional studies that do not allow causality to be addressed. Moreover, our results cannot be extrapolated to other populations, the findings of this study can only be applied to people with overweight and obesity at increased risk of CVD. In addition, this study could include the subjective decisions required in the use of cluster analysis, such as the number of clusters to implement the k-means algorithm, food group, and naming of dietary patterns. Lastly, a convenience sample was used, PREDIMED-Plus participants living in Catalonia.

5. Conclusions

This study contributes to the scarce knowledge on the relationship between the deprivation index and lifestyle in individuals with overweight/obesity and metabolic syndrome. CSI was related with lifestyle in the PREDIMED-Plus study participants living in Catalonia, Spain. Those participants with high deprivation index are at greater risk of adhering to an "unhealthy lifestyle" following an unhealthy dietary pattern and having lower physical activity. Public health policy should consider this relationship, by understanding how these factors influence lifestyle in individuals with overweight/obesity: community interventions and health policy decisions may target subsets of the population in order to promote a healthier lifestyle.

Supplementary Materials: The following are available online at https://www.mdpi.com/article/10.3390/nu13103408/s1, Table S1. Food groups. Table S2. Relationship (multivariable adjusted β-coefficients and 95%CI) between CSI and food groups consumption. Table S3. Dietary patterns and food groups.

Author Contributions: J.B., F.V. and M.P.-M. analyzed the data and wrote the article. J.B., N.B., A.G., M.D.Z., X.P., E.S. and J.S.-S. conducted and designed the research. All authors have read and agreed to the published version of the manuscript.

Funding: This work was supported by the 2017 call for proposals within the Strategic Plan of Research and Innovation in Health (PERIS) 2016–2020 for Primary Care Research Projects from the Health Department of the Generalitat de Catalunya, with reference P17/084, and the official Spanish Institutions for funding scientific biomedical research, CIBER Fisiopatología de la Obesidad y Nutrición (CIBERObn) and Instituto de Salud Carlos III (ISCIII), through the Fondo de Investigación para la Salud (FIS), which is cofunded by the European Regional Development Fund (four coordinated FIS projects led by J.S.-S., including the following projects: PI13/01123, PI13/00462, PI13/00233,

PI13/02184, PI14/00728, PI16/01094, PI16/00501, PI16/00381, PI17/00215, PI19/01226, P119/00017, PI19/00576, PI19/01032 and PI20/00339), the Especial Action Project entitled Implementación y evaluación de una intervención intensiva sobre la actividad física, Cohorte PREDIMED-Plus grant to J.S.-S.; the European Research Council (Advanced Research Grant 2014–2019; agreement #340918); the Recercaixa (number 2013ACUP00194) grant to J.S.-S. This work is partially supported by ICREA under the ICREA Academia program. None of the funding sources took part in the design, collection, analysis, interpretation of the data, or the writing of the report, or in the decision to submit the manuscript for publication.

Institutional Review Board Statement: The protocol was written in accordance with the ethical principles and good clinical practices contained in the Declaration of Helsinki. The respective Institutional Review Board of all study centers approved the study protocol.

Informed Consent Statement: All participants provided written informed consent.

Data Availability Statement: The datasets generated and analyzed during the current study are not expected to be made available outside the core research group, as neither participants' consent forms or ethics approval included permission for open access. However, the researchers will follow a controlled data sharing collaboration model, as in the informed consent participants agreed with a controlled collaboration with other investigators for research related to the project's aims. Therefore, investigators who are interested in this study can contact the PREDIMED-Plus Steering Committee by sending a request letter to predimed_plus_scommittee@googlegroups.com. A data sharing agreement indicating the characteristics of the collaboration and data management will be completed for the proposals that are approved by the Steering Committee.

Acknowledgments: The authors would especially like to thank the PREDIMED-Plus participants for their enthusiastic collaboration, the PREDIMED-Plus personnel for their outstanding support, and staff of all associated primary care centers for their exceptional work. CIBEROBN, is an initiative of the Carlos III Health Institute, Spain. We also thank the PREDIMED-Plus Biobank Network, which is part of the National Biobank Platform of the Carlos III Health Institute for storing and managing the biological samples.

Conflicts of Interest: The authors declare that they have no competing interests.

References

1. Afshin, A.; Sur, P.J.; Fay, K.A.; Cornaby, L.; Ferrara, G.; Salama, J.S.; Mullany, E.C.; Abate, K.H.; Abbafati, C.; Abebe, Z.; et al. Health effects of dietary risks in 195 countries, 1990–2017: A systematic analysis for the Global Burden of Disease Study 2017. *Lancet* **2019**, *393*, 1958–1972. [CrossRef]
2. Guthold, R.; Stevens, G.A.; Riley, L.M.; Bull, F.C. Worldwide trends in insufficient physical activity from 2001 to 2016: A pooled analysis of 358 population-based surveys with 1·9 million participants. *Lancet Glob. Health* **2018**, *6*, e1077–e1086. [CrossRef]
3. Mackenbach, J.P.; Stirbu, I.; Roskam, A.-J.R.; Schaap, M.M.; Menvielle, G.; Leinsalu, M.; Kunst, A.E. Socioeconomic Inequalities in Health in 22 European Countries. *N. Engl. J. Med.* **2008**, *358*, 2468–2481. [CrossRef] [PubMed]
4. Contoyannis, P.; Jones, A.M. Socio-economic status, health and lifestyle. *J. Health Econ.* **2004**, *23*, 965–995. [CrossRef] [PubMed]
5. Cockerham, W.C.; Abel, T.; Lüschen, G. Max Weber, Formal Rationality, and Health Lifestyles. *Sociol. Q.* **1993**, *34*, 413–428. [CrossRef]
6. Wang, J.; Geng, L. Effects of socioeconomic status on physical and psychological health: Lifestyle as a mediator. *Int. J. Environ. Res. Public Health* **2019**, *16*, 281. [CrossRef] [PubMed]
7. Groth, M.V.; Fagt, S.; Brøndsted, L. Social determinants of dietary habits in Denmark. *Eur. J. Clin. Nutr.* **2001**, *55*, 959–966. [CrossRef]
8. Martinez-Lacoba, R.; Pardo-Garcia, I.; Amo-Saus, E.; Escribano-Sotos, F. Social determinants of food group consumption based on Mediterranean diet pyramid: A cross-sectional study of university students. *PLoS ONE* **2020**, *15*, e0227620. [CrossRef] [PubMed]
9. Marques-Vidal, P.; Waeber, G.; Vollenweider, P.; Guessous, I. Socio-demographic and lifestyle determinants of dietary patterns in French-speaking Switzerland, 2009–2012. *BMC Public Health* **2018**, *18*, 131. [CrossRef]
10. Beck, K.L.; Jones, B.; Ullah, I.; McNaughton, S.A.; Haslett, S.J.; Stonehouse, W. Associations between dietary patterns, socio-demographic factors and anthropometric measurements in adult New Zealanders: An analysis of data from the 2008/09 New Zealand Adult Nutrition Survey. *Eur. J. Nutr.* **2018**, *57*, 1421–1433. [CrossRef]
11. Ciprián, D.; Navarrete-Muñoz, E.M.; Garcia de la Hera, M.; Giménez-Monzo, D.; González-Palacios, S.; Quiles, J.; Vioque, J. Patrón de dieta mediterráneo y occidental en población adulta de un área mediterránea; un análisis clúster. *Nutr. Hosp.* **2013**, *28*, 1741–1749. [PubMed]
12. Papier, K.; Jordan, S.; D'Este, C.; Banwell, C.; Yiengprugsawan, V.; Seubsman, S.A.; Sleigh, A. Social demography of transitional dietary patterns in thailand: Prospective evidence from the thai cohort study. *Nutrients* **2017**, *9*, 1173. [CrossRef] [PubMed]

13. Kesse-Guyot, E.; Bertrais, S.; Péneau, S.; Estaquio, C.; Dauchet, L.; Vergnaud, A.-C.; Czernichow, S.; Galan, P.; Hercberg, S.; Bellisle, F. Dietary patterns and their sociodemographic and behavioural correlates in French middle-aged adults from the SU.VI.MAX cohort. *Eur. J. Clin. Nutr.* **2009**, *63*, 521–528. [CrossRef] [PubMed]
14. Pérez-Tepayo, S.; Rodríguez-Ramírez, S.; Unar-Munguía, M.; Shamah-Levy, T. Trends in the dietary patterns of Mexican adults by sociodemographic characteristics. *Nutr. J.* **2020**, *19*, 1–10. [CrossRef]
15. Northstone, K.; Emmett, P.M. Dietary patterns of men in ALSPAC: Associations with socio-demographic and lifestyle characteristics, nutrient intake and comparison with women's dietary patterns. *Eur. J. Clin. Nutr.* **2010**, *64*, 978–986. [CrossRef] [PubMed]
16. Sánchez-Villegas, A.; Delgado-Rodríguez, M.; Martínez-González, M.Á.; de Irala-Estévez, J.; Martínez, J.A.; De la Fuente, C.; Alonso, A.; Guillén-Grima, F.; Aguinaga, I.; Rubio, C.; et al. Gender, age, socio-demographic and lifestyle factors associated with major dietary patterns in the Spanish project SUN (Seguimiento Universidad de Navarra). *Eur. J. Clin. Nutr.* **2003**, *57*, 285–292. [CrossRef] [PubMed]
17. Maksimov, S.; Karamnova, N.; Shalnova, S.; Drapkina, O. Sociodemographic and Regional Determinants of Dietary Patterns in Russia. *Int. J. Environ. Res. Public Health* **2020**, *17*, 328. [CrossRef] [PubMed]
18. Chastin, S.F.M.; Buck, C.; Freiberger, E.; Murphy, M.; Brug, J.; Cardon, G.; O'Donoghue, G.; Pigeot, I.; Oppert, J.-M. Systematic literature review of determinants of sedentary behaviour in older adults: A DEDIPAC study. *Int. J. Behav. Nutr. Phys. Act.* **2015**, *12*, 127. [CrossRef]
19. Xiao, Q.; Keadle, S.K.; Berrigan, D.; Matthews, C.E. A prospective investigation of neighborhood socioeconomic deprivation and physical activity and sedentary behavior in older adults. *Prev. Med.* **2018**, *111*, 14–20. [CrossRef]
20. Van Lenthe, F.J.; Brug, J.; MacKenbach, J.P. Neighbourhood inequalities in physical inactivity: The role of neighbourhood attractiveness, proximity to local facilities and safety in the Netherlands. *Soc. Sci. Med.* **2005**, *60*, 763–775. [CrossRef]
21. Darin-Mattsson, A.; Fors, S.; Kåreholt, I. Different indicators of socioeconomic status and their relative importance as determinants of health in old age. *Int. J. Equity Health* **2017**, *16*, 173. [CrossRef] [PubMed]
22. Townsend, P. Deprivation. *J. Soc. Policy* **1987**, *16*, 125–146. [CrossRef]
23. Carstairs, V.; Morris, R. Deprivation and health in Scotland. *Health Bull.* **1990**, *48*, 162–175.
24. Martinez-Beneito, M.A.; Vergara-Hernández, C.; Botella-Rocamora, P.; Corpas-Burgos, F.; Pérez-Panadés, J.; Zurriaga, Ó.; Aldasoro, E.; Borrell, C.; Cabeza, E.; Cirera, L.; et al. Geographical variability in mortality in urban areas: A joint analysis of 16 causes of death. *Int. J. Environ. Res. Public Health* **2021**, *18*, 5664. [CrossRef] [PubMed]
25. Colls, C.; Mias, M.; García-Altés, A. Un índice de privación para reformar el modelo de financiación de la atención primaria en Cataluña. *Gac. Sanit.* **2020**, *34*, 44–50. [CrossRef] [PubMed]
26. Garciá-Altés, A.; Ruiz-Munõz, D.; Colls, C.; Mias, M.; Martín Bassols, N. Socioeconomic inequalities in health and the use of healthcare services in Catalonia: Analysis of the individual data of 7.5 million residents. *J. Epidemiol. Community Health* **2018**, *72*, 871–879. [CrossRef] [PubMed]
27. Black, J.L.; Macinko, J. Neighborhoods and obesity. *Nutr. Rev.* **2008**, *66*, 2–20. [CrossRef]
28. Martínez-González, M.A.; Buil-Cosiales, P.; Corella, D.; Bulló, M.; Fitó, M.; Vioque, J.; Romaguera, D.; Martínez, J.A.; Wärnberg, J.; López-Miranda, J.; et al. Cohort Profile: Design and methods of the PREDIMED-Plus randomized trial. *Int. J. Epidemiol.* **2018**, *48*, 387–388o. [CrossRef]
29. Willett, W. *Nutritional Epidemiology*, 3rd ed.; United States of America by Oxford University Press: New York, NY, USA, 2013.
30. Fernández-Ballart, J.D.; Piñol, J.L.; Zazpe, I.; Corella, D.; Carrasco, P.; Toledo, E.; Perez-Bauer, M.; Martínez-González, M.Á.; Salas-Salvadó, J.; Martn-Moreno, J.M. Relative validity of a semi-quantitative food-frequency questionnaire in an elderly Mediterranean population of Spain. *Br. J. Nutr.* **2010**, *103*, 1808–1816. [CrossRef]
31. Schröder, H.; Zomeño, M.D.; Martínez-González, M.A.; Salas-Salvadó, J.; Corella, D.; Vioque, J.; Romaguera, D.; Martínez, J.A.; Tinahones, F.J.; Miranda, J.L.; et al. Validity of the energy-restricted Mediterranean Diet Adherence Screener. *Clin. Nutr.* **2021**, *40*, 4971–4979. [CrossRef]
32. Schröder, H.; Corella, D.; Salas-salvado, J.; Lamuela-ravento, R.; Ros, E.; Salaverrı, I.; Vinyoles, E.; Go, E. A Short Screener Is Valid for Assessing Mediterranean Diet Adherence among Older. *J. Nutr.* **2011**, *141*, 1140–1145. [CrossRef] [PubMed]
33. Elosua, R.; Marrugat, J.; Molina, L.; Pons, S.; Pujol, E. Validation of the Minnesota Leisure Time Physical Activity Questionnaire in Spanish men. The MARATHOM Investigators. *Am. J. Epidemiol.* **1994**, *139*, 1197–1209. [CrossRef] [PubMed]
34. Elosua, R.; Garcia, M.; Aguilar, A.; Molina, L.; Covas, M.I.; Marrugat, J. Validation of the Minnesota leisure time physical activity questionnaire in Spanish women. Investigators of the MARATDON Group. *Med. Sci. Sports Exerc.* **2000**, *32*, 1431–1437. [CrossRef]
35. Martínez-González, M.A.; López-Fontana, C.; Varo, J.J.; Sánchez-Villegas, A.; Martinez, J.A. Validation of the Spanish version of the physical activity questionnaire used in the Nurses' Health Study and the Health Professionals' Follow-up Study. *Public Health Nutr.* **2005**, *8*, 920–927. [CrossRef]
36. Willett, W.; Stampfer, M.J. Total energy intake: Implications of total energy intake for epidemiologic Analyses. *Am. J. Epidemiol.* **1986**, *124*, 273–301. [CrossRef]
37. Ochieng, B.M.N. Factors affecting choice of a healthy lifestyle: Implications for nurses. *Br. J. Community Nurs.* **2006**, *11*, 78–81. [CrossRef]

38. Algren, M.H.; Bak, C.K.; Berg-Beckhoff, G.; Andersen, P.T. Health-Risk Behaviour in Deprived Neighbourhoods Compared with Non-Deprived Neighbourhoods: A Systematic Literature Review of Quantitative Observational Studies. *PLoS ONE* **2015**, *10*, e0139297. [CrossRef]
39. Mumme, K.; Conlon, C.; von Hurst, P.; Jones, B.; Stonehouse, W.; Heath, A.-L.M.; Coad, J.; Haskell-Ramsay, C.; de Seymour, J.; Beck, K. Dietary Patterns, Their Nutrients, and Associations with Socio-Demographic and Lifestyle Factors in Older New Zealand Adults. *Nutrients* **2020**, *12*, 3425. [CrossRef]
40. Kurotani, K.; Honjo, K.; Nakaya, T.; Ikeda, A.; Mizoue, T.; Sawada, N.; Tsugane, S. Diet quality affects the association between census-based neighborhood deprivation and all-cause mortality in japanese men and women: The Japan public health center-based prospective study. *Nutrients* **2019**, *11*, 2194. [CrossRef]
41. Foster, H.M.E.; Celis-Morales, C.A.; Nicholl, B.I.; Petermann-Rocha, F.; Pell, J.P.; Gill, J.M.R.; O'Donnell, C.A.; Mair, F.S. The effect of socioeconomic deprivation on the association between an extended measurement of unhealthy lifestyle factors and health outcomes: A prospective analysis of the UK Biobank cohort. *Lancet Public Health* **2018**, *3*, e576–e585. [CrossRef]
42. Lakshman, R.; McConville, A.; How, S.; Flowers, J.; Wareham, N.; Cosford, P. Association between area-level socioeconomic deprivation and a cluster of behavioural risk factors: Cross-sectional, population-based study. *J. Public Health* **2011**, *33*, 234–245. [CrossRef] [PubMed]
43. Beaulac, J.; Kristjansson, E.; Cummins, S. A systematic review of food deserts, 1966-2007. *Prev. Chronic Dis.* **2009**, *6*, A105. [PubMed]
44. Caspi, C.E.; Sorensen, G.; Subramanian, S.V.; Kawachi, I. The local food environment and diet: A systematic review. *Health Place* **2012**, *18*, 1172–1187. [CrossRef] [PubMed]
45. Lachowycz, K.; Jones, A.P. Does walking explain associations between access to greenspace andlower mortality? *Soc. Sci. Med.* **2014**, *107*, 9–17. [CrossRef] [PubMed]
46. Jones, A.; Hillsdon, M.; Coombes, E. Greenspace access, use, and physical activity: Understanding the effects of area deprivation. *Prev. Med.* **2009**, *49*, 500–505. [CrossRef] [PubMed]
47. Ek, S. Gender differences in health information behaviour: A Finnish population-based survey. *Health Promot. Int.* **2015**, *30*, 736–745. [CrossRef]
48. Wardle, J.; Haase, A.M.; Steptoe, A.; Nillapun, M.; Jonwutiwes, K.; Bellisle, F. Gender Differences in Food Choice: The Contribution of Health Beliefs and Dieting. *Ann. Behav. Med.* **2004**, *27*, 107–116. [CrossRef]

MDPI

Article

Adherence to the Mediterranean Diet in Association with Self-Perception of Diet Sustainability, Anthropometric and Sociodemographic Factors: A Cross-Sectional Study in Italian Adults

Beatrice Biasini [1], Alice Rosi [1], Davide Menozzi [2] and Francesca Scazzina [1,*]

[1] Department of Food and Drug, University of Parma, Via Volturno 39, 43125 Parma, Italy; beatrice.biasini@unipr.it (B.B.); alice.rosi@unipr.it (A.R.)
[2] Department of Food and Drug, University of Parma, Via Kennedy 6, 43125 Parma, Italy; davide.menozzi@unipr.it
* Correspondence: francesca.scazzina@unipr.it; Tel.: +39-0521-906203

Abstract: The adoption of sustainable dietary models, such as the Mediterranean Diet (MD), can be a valuable strategy to preserve ecosystems and human health. This study aims to investigate in an Italian adult representative sample the adherence to the MD and to what extent it is associated with the self-perceived adoption of a sustainable diet, the consideration of the MD as a sustainable dietary model, and anthropometric and sociodemographic factors. By applying an online survey (n = 838, 18–65 years, 52% female), an intermediate level of MD adherence (median: 4.0, IR: 3.0–4.0) in a 0–9 range was observed. Only 50% of the total sample confirmed the MD as a sustainable dietary model, and 84% declared no or low perception of adopting a sustainable diet. Being female, having a higher income and education level, considering the MD as a sustainable dietary model, as well as the perception of having a sustainable diet were the most relevant factors influencing the probability of having a high score (≥6) of adherence to the MD. This study suggests a gradual shift away from the MD in Italy and supports the need to address efforts for developing intervention strategies tailored to adults for improving diet quality. Furthermore, a public campaign should stress the link between a diet and its environmental impact to foster nutritionally adequate and eco-friendly dietary behaviors.

Keywords: Mediterranean Diet; sustainable diet; food frequency questionnaire; diet self-perception; socioeconomic profile; health status; adult population

Citation: Biasini, B.; Rosi, A.; Menozzi, D.; Scazzina, F. Adherence to the Mediterranean Diet in Association with Self-Perception of Diet Sustainability, Anthropometric and Sociodemographic Factors: A Cross-Sectional Study in Italian Adults. *Nutrients* **2021**, *13*, 3282. https://doi.org/10.3390/nu13093282

Academic Editors: Daniela Bonofiglio and Maria D. Mesa

Received: 29 July 2021
Accepted: 17 September 2021
Published: 20 September 2021

Publisher's Note: MDPI stays neutral with regard to jurisdictional claims in published maps and institutional affiliations.

1. Introduction

Dietary patterns and the frequency of food consumption are associated with a person's health status, but they also substantially modulate resource exploitation. It has been largely proven that several health advantages can be linked to the adoption of a dietary pattern inspired by the Mediterranean Diet (MD) [1,2]. A Mediterranean-type diet is able to prevent the development of cardiovascular disease—not only in populations living in the Mediterranean area [1,3,4]—reducing the risk of diabetes [5] and metabolic syndrome [6,7]. In addition, an inverse association between adherence to the MD and the risk of several cancer types and cancer mortality [8], depression, and cognitive impairment [9] has been demonstrated, with beneficial effects on sleep quality in the adult population [10].

The MD is a nutritionally adequate plant-centered diet whose pillars are similar to those qualifying the global "healthy diet from sustainable food systems" described by the EAT Lancet Commission [11]. It is grounded in the "One Health" concept, strengthening the idea that human and ecosystem health are not independent [12,13]. Both dietary approaches largely include whole grains, fruit and vegetables, nuts, and unsaturated fatty acid sources (e.g., olive oil), and both limit the consumption of red meat, processed meat, starchy vegetables, added sugar, and refined grains. However, a stricter limitation is addressed to fish and white meat intake in the reference diet described by the EAT Lancet Commission [11,14]. If food security, affordability, accessibility, and cultural acceptability are ensured, the MD emerges as an example of a sustainable dietary pattern able to address health and ecological concerns [15,16], showing a lower environmental impact, richness in biodiversity, important food-related sociocultural values, and positive economic return for local communities [17]. As found in previous studies, the dietary shift toward a Mediterranean-type diet can lead to positive outcomes both for health and climate change [18,19].

It is worth noting that the MD, acknowledged by UNESCO as an intangible cultural heritage of humanity, should not be considered just as a mere reference from which a particular set of food needs to be consumed in specified quantities and proportions, but also as a cultural model involving not only consumption, but food production, processing, distribution, and cooking, including rituals and traditions [14,20]. On the basis of the health-promoting outcomes linked to the adoption of a Mediterranean dietary pattern, scientific efforts have been increasingly addressed to develop proper methods to estimate the adherence to the MD by using indexes or scores and to test for their disease risk predictive ability [21]. As a result, such indexes or scores can be calculated from the intake of certain food components and lifestyle factors [22].

In the last few decades, Mediterranean populations, including Italian adolescents [23] and adults [24], are stepping away from their traditional dietary patterns, such as the MD. Important determinants are globalization, which overcomes geographical barriers between food productions and consumption worldwide, population growth, urbanization, and lifestyle changes, together with economic and sociocultural factors [15,25]. Food system and food consumption pattern transformations are responsible for a loss of biodiversity and soil degradation and represent significant challenges for the state of food security and nutrition [26].

Based on these considerations, the aim of the present study was to investigate the adherence to the MD and to what extent this variable is associated with the self-perception of the MD's sustainability, as well as the adoption of a sustainable diet in a representative sample of Italian adults. The secondary aim was to explore how demographic, socioeconomic, and behavioral variables may influence adherence to the MD.

2. Materials and Methods

2.1. Sample

After receiving approval from the local institutional ethical committee (Comitato Etico Area Vasta Emilia Nord, 1139/2018/OSS/UNIPR), an online survey instrument was distributed to a representative sample of adults residing in Italy (18–65 years) through a software platform from a marketing agency (Qualtrics International Inc., Seattle, Washington and Provo, Utah, United States of America) in July 2019. The agency invited subjects to participate by sending communication via e-email to pre-enrolled members. The survey, which is part of a PhD thesis [27], was completed on one occasion by the respondents, who provided their informed consent to the study. Only the subjects living in Italy and not affected by cardiometabolic conditions, such as diabetes and cardiovascular diseases, as well as eating disorders were eligible to participate in the present cross-sectional study. The data were collected from subjects representative of the adult population residing in Italy based on three selected criteria: gender distribution, age range, and geographical areas of residence (nomenclature of territorial units for statistic (NUTS 1)). When a sufficient number of individuals completed the survey, subject recruitment and online survey self-administration was stopped. The enrolled participants received compensation after the survey's completion. To be representative of the entire Italian adult population (n = 37,248,990, as indicated by the dataset provided by the National Institute of Statistics (ISTAT), referred to on 1 January 2019), the minimum sample size was set at 666 participants, taking into account a confidence level of 99% and confidence interval of 5%. As some participants could drop out during the study or could be excluded from the analyses because of missing answers, more than 800 respondents were invited to participate to the study. The calculation of the sample size was performed using the sample size calculator suggested by the National Institute of Health (https://www.epicentro.iss.it/strumenti/SampleSize, accessed on 30 June 2021).

2.2. Measures

Anthropometric and sociodemographic information was self-reported by the subjects. Height and weight were assessed as continuous variables, while others were assessed as categorical variables, including sex and nationality (2 categories each), educational attainment and income (3 categories each), size of residence, number of members and children within the household (4 categories each), age, occupation, and geographical area of residence (5 categories each). Using weight and height data, the subjects' BMIs were computed, and weight status was defined by applying the WHO's standard cut-offs [28]. The subjects were asked to express the degree of responsibility for food purchasing and meal preparation (e.g., being the main person responsible, co-responsible, or little or not at all responsible), the habitual frequency of eating out (e.g., never or seldom, <1 time/week, 1 time/week, 2–4 times/week, or ≥5 times/week), the presence of a certain physiological status (e.g., pregnancy or breastfeeding), particular health risk factors (e.g., hypertension or dyslipidemia), and food allergies or intolerances. In addition, the respondents were asked to indicate any participation in environmental associations (EAs) or solidarity purchasing groups (SPGs), which are local networks of people who organize collective purchase decisions regarding food and other goods, selecting suppliers based on critical consumption and solidarity criteria. Specifically, SPGs are intended to promote environmental sustainability (e.g., selecting seasonal, organic, or local products) and social sustainability with respect to the producers (by creating social bonds with them) and SPG members themselves (e.g., by providing mutual assistance) [29,30].

Adherence to the MD was assessed by using a 15-item food frequency questionnaire, already validated to determine the adherence to the Mediterranean dietary pattern of Italian adults on a score from 0 (minimal adherence) to 9 (maximal adherence) [31]. The scoring scheme of the components was binary, with a score of 0 or 1 being associated to the intake of each included food group or item, expressed as the number of portions per day or week. In detail, 1 point was assigned as follows: vegetables (≥2/day), fresh fruit (≥2/day), dried fruit (≥2/week), wholegrain cereals (≥1/day), pulses (≥2/week), fish (≥2/week), olive oil (≥3/day), red and processed meat (<1–3/week), and wine (1–2 glasses per day for men and less than 1 glass per day for women). The final score was computed by summing each individual score. In addition, according to the obtained level of adherence (tertiles) to the MD, the respondents were divided into low (fist tertile, MD score 0–2), medium (second tertile, MD score 3–5) and high (third tertile, MD score 6–9) adherence groups. Moreover, the level of adherence to the MD was evaluated according to the subjects' compliance with the Italian national recommended intake for the food groups or items considered for a standard dietary energy intake of 2000 kcal/day [32]: fruits (3/day), vegetables (≥2/day), nuts (2/week), legumes (3/week), red meat (≤1/week), fish (2–3/week), wine (never or hardly ever), white meat (2/week), sweets (<1/week), butter, margarine, or cooking cream (≤3/day), olive oil (3/day), milk or yogurt (3/day), and carbonated or sugar-sweetened beverages (<1/week).

After providing the respondents with the definition of sustainable diets expressed by the Food and Agricultural Organization (FAO) [33], including a more explicit and concrete description to provide respondents with a unique interpretation, the online survey also included a question addressed to understanding if the respondents perceived the MD as a sustainable diet (i.e., "Do you think that the MD can be considered a sustainable dietary model?"). This item was adapted from those developed by Riddell et al. [34] for measuring the self-perception of diet and healthy eating. In addition, the self-perceived adoption of a sustainable diet during the last 3 months was assessed (i.e., "I can say that I have adopted a sustainable diet within the last 3 months") as adapted from Fishbein and Ajzen [35].

Both the answers were measured on a unipolar 7-point scale anchored by totally disagree or totally agree answers. However, for reasons pertaining to result interpretation, the answers were collapsed into 3 categories: "no", including those who *disagreed* or *totally disagreed*; "not much/maybe", corresponding to those who *somewhat disagreed*, were *neutral*, or *somewhat agreed*; and "yes", referring to those who *agreed* or *strongly agreed*.

2.3. Data Analysis

Descriptive and inferential statistics were collected. Normality of the data distribution was rejected through the Kolmogorov–Smirnov test. The results were expressed as a frequency (%) or as median and interquartile ranges. The Chi-square test (χ^2) was used to explore potential associations between gender and (1) demographic and socioeconomic characteristics, (2) adherence to the MD, (3) the perception of the MD as a sustainable dietary model, and (4) self-perceived adoption of a sustainable diet. The same test was applied also to investigate potential associations between adherence to the MD (with the subjects divided in MD score-based tertiles) and frequency of food consumption recommended by the Italian dietary guidelines [32]. The non-parametric Mann–Whitney test was applied to explore differences between genders (males and females). Furthermore, to evaluate which characteristics were able to predict scores of high adherence to the MD, univariate and multivariate logistic regression analyses were carried out. A p-value less than 0.05 was considered statistically significant. All statistical analyses were performed with IBM SPSS Statistics for Windows, version 25.0 (Armonk, NY: IBM Corp).

3. Results

3.1. Participants' Characteristics and Adherence to the MD

Overall, 860 subjects answered the online survey. After removing the low-quality records ($n = 22$), the final sample was composed of 838 respondents, representative of the adult residents in Italy. The participants' characteristics are provided in Table 1. Approximately half of the respondents were females (52%), and most of the respondents were from 35 to 65 years old (71%). The vast majority of the sample attained at least secondary education (79%). Different employment conditions were observed by comparing males and females, with more than two-thirds of males working as full-time employees and more than one-fourth of females being unemployed. Most of the respondents were apparently healthy, declaring a normal body weight; however, a significant association was found when the sample split by gender was grouped by BMI categories, with a higher proportion of males being overweight or obese compared with females ($p < 0.001$). Moreover, a higher percentage of females compared with males had the main responsibility of food purchasing and meal preparation.

After dividing the respondents according to their level of adherence to the MD, an association between gender and the level of adherence to the MD was spotted ($p < 0.001$). A medium compliance to the MD was reported, with females presenting significantly higher scores compared with males ($p < 0.001$).

Although only 1% of the subjects clearly disagreed in considering the MD as sustainable, half of the sample (49%) expressed uncertainty with this statement. On the other hand, a low number of respondents (16%) (strongly) agreed to having adopted a sustainable diet in the last 3 months.

3.2. Compliance with Food Recommendations

Table 2 shows the number and proportions of respondents being compliant with the consumption frequency or frequencies used to compute the MD score or those recommended by Italian guidelines. Significant associations were found between the MD adherence categories (low, medium, and high) and all the food items or groups used to compute the final MD. In addition, the level of adherence to the MD was found to be in association with being compliant with the Italian national recommended intake for fruit, vegetables, milk and yogurt, red meat, carbonated or sweet beverages, fish and seafood, nuts, and pulses.

Table 1. Anthropometric variables, demographic and socioeconomic characteristics, health conditions, food-related habits, adherence to the MD, consideration of the MD as a sustainable dietary model, and self-perceived adoption of a sustainable diet in the last 3 months in the total sample, split by gender.

	All (n = 838)	Female (n = 434)	Male (n = 404)	p Value
Age range (years)				0.131 [a]
18–24	89 (10.6)	53 (12.2)	36 (8.9)	
25–34	157 (18.7)	79 (18.2)	78 (19.3)	
35–44	198 (23.6)	111 (25.6)	87 (21.5)	
45–54	209 (24.9)	95 (21.9)	114 (28.2)	
55–65	185 (22.1)	96 (22.1)	89 (22.0)	
BMI (kg/m^2)				<0.001 [a]
<18.5 (underweight)	36 (4.3)	33 (7.6)	3 (0.7)	
18.5–24.9 (normal weight)	497 (59.3)	284 (65.4)	213 (52.7)	
25.0–29.9 (overweight)	230 (27.4)	82 (18.9)	148 (36.6)	
≥30.0–34.9 (obesity)	75 (8.9)	26 (6.0)	31 (7.7)	
Health conditions				<0.001 [a]
Anemia, hypertension, or dyslipidemia	83 (9.9)	36 (8.3)	47 (11.6)	
Food intolerance or allergies	138 (16.5)	70 (16.1)	68 (16.8)	
Menopause, pregnancy, or breastfeeding	70 (8.4)	70 (16.1)	-	
None of the above	547 (65.3)	258 (59.4)	289 (71.5)	
Geographical area of residence				0.683 [a]
Northwest	220 (26.3)	116 (26.7)	104 (25.7)	
Northeast	168 (20.0)	86 (19.8)	82 (20.3)	
Central	167 (19.9)	85 (19.6)	82 (20.3)	
South	192 (22.9)	94 (21.7)	98 (24.3)	
Islands	91 (10.9)	53 (12.2)	38 (9.4)	
Size of residence (number of inhabitants)				0.300 [a]
<5000	148 (17.7)	80 (18.4)	68 (16.8)	
5000–49,999	348 (41.5)	188 (43.3)	160 (39.6)	
50,000–500,000	206 (24.6)	95 (21.9)	111 (27.5)	
>500,000	136 (16.2)	71 (16.4)	65 (16.1)	
Education level				0.151 [a]
Primary or lower secondary	62 (7.4)	26 (6.0)	36 (8.9)	
Secondary	448 (53.5)	243 (56.0)	205 (50.7)	
Tertiary *	328 (39.1)	165 (38.0)	163 (40.3)	
Occupation				<0.001 [a]
Full-time employee	441 (52.6)	167 (38.5)	274 (67.8)	
Part-time employee	133 (15.9)	93 (21.4)	40 (9.9)	
Unemployed	158 (18.9)	113 (26.0)	45 (11.1)	
Retired	35 (4.2)	16 (3.7)	19 (4.7)	
Student	71 (8.5)	45 (10.4)	26 (6.4)	
Monthly household net income				0.005 [a]
≤EUR 1499	193 (23.0)	115 (26.5)	78 (19.3)	
EUR 1500–2499	267 (31.9)	132 (30.4)	135 (33.4)	
≥EUR 2500	294 (35.1)	131 (30.2)	163 (40.3)	
Do not wish to tell or do not know	84 (10.0)	56 (12.9)	28 (6.9)	
N. household members				0.084 [a]
1	76 (9.1)	33 (7.6)	43 (10.6)	
2	182 (21.7)	106 (24.4)	76 (18.8)	
3	267 (31.9)	129 (29.7)	138 (34.2)	
>3	313 (37.4)	166 (38.2)	147 (36.4)	

Table 1. *Cont.*

	All (*n* = 838)	Female (*n* = 434)	Male (*n* = 404)	*p* Value
N. household members < 18 years				0.263 [a]
None	526 (62.8)	284 (65.4)	242 (59.9)	
1	172 (20.5)	80 (18.4)	92 (22.8)	
2	118 (14.1)	57 (13.1)	61 (15.1)	
≥3	22 (2.6)	13 (3.0)	9 (2.2)	
Responsibility of food purchases				<0.001 [a]
Mainly responsible	601 (71.7)	347 (80.0)	254 (62.9)	
Co-responsible	222 (26.5)	84 (19.4)	138 (34.2)	
Little or not at all responsible	15 (1.8)	3 (0.7)	12 (3.0)	
Responsibility in meal preparation				<0.001 [a]
Mainly responsible	529 (63.1)	346 (79.7)	183 (45.3)	
Co-responsible	262 (31.3)	78 (18.0)	184 (45.5)	
Little or not at all responsible	47 (5.6)	10 (2.3)	37 (9.2)	
Frequency of eating out				<0.001 [a]
Never or seldom	129 (15.4)	83 (19.1)	46 (11.4)	
<1 time/week	167 (19.9)	101 (23.3)	66 (16.3)	
1 time/week	195 (23.3)	101 (23.3)	94 (23.3)	
2–4 times/week	249 (29.7)	117 (27.0)	132 (23.7)	
≥5 times/week	98 (11.7)	32 (7.4)	66 (16.3)	
Taking part in SPGs or EAs				0.634 [a]
Yes	93 (11.1)	46 (10.6)	47 (11.6)	
No	745 (88.9)	388 (89.4)	357 (88.4)	
MD score (on a 0–9 scale)	4.0 (3.0–5.0)	4.0 (3.0–6.0)	3.0 (2.0–5.0)	<0.001 [b]
Adherence to the Mediterranean Diet				<0.001 [a]
Low	164 (19.6)	52 (12.0)	112 (27.7)	
Medium	498 (59.4)	261 (60.1)	237 (58.7)	
High	176 (21.0)	121 (27.9)	55 (13.6)	
MD considered a sustainable dietary model				0.375 [a]
No	10 (1.2)	3 (0.7)	7 (1.7)	
Maybe	410 (48.9)	212 (48.8)	198 (49.0)	
Yes	418 (49.9)	219 (50.5)	199 (49.3)	
Self-perceived adoption of a sustainable diet within the last 3 months				0.317 [a]
No	175 (20.9)	85 (19.6)	90 (22.3)	
Not much	526 (62.8)	283 (65.2)	243 (60.1)	
Yes	137 (16.3)	66 (15.2)	71 (17.6)	

Note: * including short cycle tertiary education. Data are expressed as a number (%) or as the median (IR). [a] Chi-square test. [b] Non-parametric Mann–Whitney test for independent sample. EAs: environmental associations; SPGs: solidarity purchasing groups.

Irrespective of the level of adherence to the MD, the food frequency questionnaire provided a qualitative descriptive picture of respondents' dietary habits (Supplementary Table S1). More than half of the subjects indicated they ate one or less than one portion per day of fruits (59%) or vegetables (69%). Similarly, 61% and 55% of respondents stated they consumed less than one portion per day of wholegrain pasta or rice as well as bread or its substitutes, respectively, with approximately 20% of the whole sample never or hardly ever eating wholegrain products. More than 50% of people stated that they instead consumed from 1 to 3 portions a week of red or white meat, while 9% consumed 4 or more portions of red meat per week. Furthermore, 72% of the respondents consumed no more than 1 portion per week of legumes.

Table 2. Compliance with food consumption recommendations used to compute the MD score (MD) and following the Italian guidelines (IT) for healthy eating for a standard dietary energy intake of 2000 kcal/day, according to the level of adherence to the MD.

	Ref. Intake	Adherence to the MD			p Value
		Low (n = 164)	**Medium (n = 498)**	**High (n = 176)**	
Wholegrains	MD (≥1/d)	87 (53.0)	417 (83.7)	158 (89.8)	<0.001
	IT (n.a.)	n.a.	n.a.	n.a.	
Vegetables	MD (≥2/d)	3 (1.8)	121 (24.3)	137 (77.8)	<0.001
	IT (≥2/d)	3 (1.8)	121 (24.3)	137 (77.8)	<0.001
Fruit	MD (≥2/d)	8 (4.9)	188 (37.8)	150 (85.2)	<0.001
	IT (≥3/d)	–	33 (6.6)	41 (23.3)	<0.001
Milk and yogurt	MD (n.a.)	n.a.	n.a.	n.a.	
	IT (≥3/d)	1 (0.6)	8 (1.6)	8 (4.5)	0.021
Olive oil	MD (≥3/d)	1 (0.6)	46 (9.2)	64 (36.4)	<0.001
	IT (3–4/d)	1 (0.6)	41 (8.2)	54 (30.7)	<0.001
Butter, margarine, or cooking cream	MD (n.a.)	n.a.	n.a.	n.a.	
	IT (<3/d)	163 (99.4)	488 (98.0)	170 (96.6)	0.187
Wine	MD (1–2/d, M; >0 < 1/d, F)	44 (26.8)	291 (58.4)	134 (76.1)	<0.001
	IT (never or hardly never)	81 (49.4)	236 (47.4)	88 (50.0)	0.800
Red meat or meat products	MD (≤3/w)	125 (76.2)	461 (92.6)	173 (98.3)	<0.001
	IT (≤1/w)	53 (32.3)	183 (36.7)	94 (53.4)	<0.001
White meat	MD (n.a.)	n.a.	n.a.	n.a.	
	IT (1–3/w)	84 (51.2)	296 (59.4)	97 (55.1)	0.158
Carbonated or SSB	MD	n.a.	n.a.	n.a.	
	IT (<1/w)	108 (65.9)	346 (69.5)	142 (80.7)	0.005
Sweets	MD	n.a.	n.a.	n.a.	
	IT (<1/w)	103 (62.8)	295 (59.2)	112 (63.6)	0.501
Fish or seafood	MD (≥2/w)	4 (2.4)	155 (31.1)	115 (65.3)	<0.001
	IT (2–3/w)	3 (1.8)	146 (29.3)	104 (59.1)	<0.001
Nuts	MD (≥2/w)	3 (1.8)	123 (24.7)	117 (66.5)	<0.001
	IT (2–3/w)	2 (1.2)	89 (17.9)	70 (39.8)	<0.001
Pulses	MD (≥2/w)	7 (4.3)	120 (24.1)	111 (63.1)	<0.001
	IT (2–3/w)	7 (4.3)	112 (22.5)	96 (54.5)	<0.001

Note: data are expressed as a number (%). The recommended food consumption frequency on a daily (d) or weekly (w) basis of the reference serving(s) are shown in the second column. The reported national reference intake for fruit and milk or yogurt (≥3 instead of 3 servings/day), olive oil (3–4 instead of 3 servings/day) butter, margarine, or cooking cream (<3 instead of ≤3 servings/day), white meat (1–3 instead of 2 servings/day), pulses (2–3 instead of 3 servings/week), and nuts (2–3 instead of 2 servings/week) have been slightly adapted to equal the categorization of the food consumption frequency provided by the MD questionnaire. Chi-square test. Low = first tertile; Medium = second tertile; High = third tertile; SSB: sugar-sweetened beverages.

3.3. *Associations between Adherence to the MD and Anthropometric, Sociodemographic, and Sustainability Perception of Diet Variables*

Among the factors influencing the probability of having a high score of adherence to the MD, being female, having a higher income and educational level, considering the MD a sustainable dietary model, as well as perceiving having a sustainable diet were the most relevant, as they were found to be statistically significant in the univariate and multivariate regression analysis (Table 3). Other conditions such as being overweight or obese and having no or little responsibility in food purchases or meal preparation significantly decreased the probability of having a high adherence level to the MD only when tested singularly in the univariate analysis, suggesting its being less incisive in affecting this outcome. An opposite effect instead was found for pregnancy, breastfeeding, experiencing menopause, and living in a big city (n. of inhabitants > 500,000), which positively influenced the adherence score, according to the same analysis.

Table 3. Logistic regression analysis for being in the third tertile of distribution of the adherence score to the MD (6-9 points) by considering all the assessed variables alone (univariate analysis) or together (multivariate analysis).

Variables	Univariate Analysis		Multivariate Analysis	
	OR (95% CI)	*p* Value	OR (95% CI)	*p* Value
Gender				
Females	-1-		-1-	
Males	0.408 (0.286–0.580)	<0.001	0.366 (0.220–0.609)	<0.001
Age (years)				
18–24	-1-		-1-	
25–34	0.948 (0.479–1.873)	0.877	0.459 (0.164–1.287)	0.139
35–44	1.342 (0.711–2.532)	0.364	0.704 (0.251–1.979)	0.506
45–54	1.182 (0.625–2.234)	0.607	0.764 (0.275–2.123)	0.606
55–65	1.467 (0.776–2.772)	0.239	0.667 (0.222–2.002)	0.470
BMI (kg/m^2)				
<18.5	-1-		-1-	
18.5–24.9	0.623 (0.302–1.283)	0.199	0.693 (0.265–1.811)	0.454
25.0–29.9	0.383 (0.176–0.834)	0.016	0.632 (0.225–1.774)	0.383
≥30.0	0.273 (0.102–0.728)	0.010	0.647 (0.187–2.235)	0.491
Health conditions				
Anemia, hypertension, or dyslipidemia	1.180 (0.663–2.100)	0.573	1.062 (0.503–2.242)	0.875
Food intolerance or allergies	1.498 (0.960–2.336)	0.075	1.558 (0.909–2.669)	0.106
Menopause, pregnancy, or breastfeeding	2.877 (1.696–4.880)	<0.001	1.924 (0.916–4.042)	0.084
None of the above	-1-		-1-	
Geographical area of residence				
Northwest	-1-		-1-	
Northeast	0.986 (0.608–1.599)	0.953	1.008 (0.537–1.893)	0.980
Central	0.925 (0.567–1.510)	0.756	0.840 (0.457–1.544)	0.574
South	0.918 (0.573–1.471)	0.723	1.109 (0.605–2.033)	0.738
Islands	0.689 (0.364–1.304)	0.252	0.635 (0.280–1.438)	0.276
Size of residence (number of inhabitants)				
<5000	-1-		-1-	
5000–49,999	1.160 (0.705–1.911)	0.559	0.880 (0.471–1.643)	0.687
50,000–500,000	1.166 (0.677–2.009)	0.580	0.876 (0.439–1.749)	0.707
>500,000	1.955 (1.115–3.428)	0.019	1.622 (0.791–3.324)	0.187
Educational level				
Primary or lower secondary	-1-		-1-	
Secondary	2.946 (1.148–7.560)	0.025	3.098 (1.020–9.410)	0.046
Tertiary * or higher	3.617 (1.401–9.339)	0.008	3.072 (0.973–9.700)	0.056
Occupation				
Full-time employee	-1-		-1-	
Part-time employee	1.186 (0.753–1.867)	0.461	1.372 (0.737–2.551)	0.318
Unemployed	0.741 (0.462–1.187)	0.213	1.041 (0.514–2.107)	0.912
Retired	1.437 (0.667–3.097)	0.354	1.961 (0.706–5.451)	0.196
Student	0.589 (0.291–1.193)	0.142	0.599 (0.181–1.982)	0.401
Monthly household net income				
≤EUR 1499	-1-		-1-	
EUR 1500–2499	1.705 (1.026–2.832)	0.039	1.950 (1.051–3.620)	0.034
≥EUR 2500	2.160 (1.324–3.527)	0.002	2.419 (1.225–4.777)	0.011
Number of household members				
1	-1-		-1-	
2	1.195 (0.594–2.405)	0.618	0.890 (0.364–2.178)	0.798
3	1.623 (0.841–3.135)	0.149	1.880 (0.748–4.725)	0.180
>3	1.149 (0.594–2.224)	0.679	0.722 (0.261–1.995)	0.529

Table 3. *Cont.*

Variables	Univariate Analysis		Multivariate Analysis	
	OR (95% CI)	*p* Value	OR (95% CI)	*p* Value
Number of household members <18 years				
None	-1-		-1-	
1	1.179 (0.774–1.794)	0.443	0.576 (0.312–1.065)	0.079
2	1.417 (0.888–2.261)	0.144	1.789 (0.793–4.033)	0.161
≥3	1.559 (0.595–4.083)	0.366	1.970 (0.549–7.060)	0.298
Responsibility of food purchases #				
Main responsible	-1-		-1-	
Co-, little, or not at all responsible	0.495 (0.326–0.752)	0.001	0.905 (0.446–1.836)	0.781
Responsibility in food preparation				
Main responsible	-1-		-1-	
Co-responsible	0.516 (0.348–0.763)	0.001	0.875 (0.453–1.690)	0.690
Little or not at all responsible	0.201 (0.061–0.658)	0.008	0.748 (0.170–3.287)	0.700
Frequency of eating out				
Never or seldom	-1-		-1-	
<1 time/week	1.648 (0.897–3.029)	0.108	1.730 (0.815–3.674)	0.154
1 time/week	1.737 (0.963–3.133)	0.067	1.906 (0.903–4.022)	0.091
2–4 times/week	1.641 (0.927–2.907)	0.089	1.405 (0.649–3.039)	0.388
≥5 times/week	1.484 (0.743–2.965)	0.263	1.317 (0.534–3.252)	0.550
Taking part in SPGs or EAs				
Yes	-1-		-1-	
No	0.650 (0.399–1.057)	0.083	1.038 (0.557–1.936)	0.906
MD considered a sustainable dietary model #				
No/maybe	-1-		-1-	
Yes	2.617 (1.840–3.723)	<0.001	2.293 (1.487–3.534)	<0.001
Self-perceived adoption of a sustainable diet				
No	-1-		-1-	
Not much	3.388 (1.815–6.323)	<0.001	2.162 (1.089–4.293)	0.028
Yes	10.275 (5.222–20.216)	<0.001	7.667 (3.517–16.711)	<0.001

Note: * including short cycle tertiary education. # The categories "No" and "Maybe", as well as the categories "Co-responsible" and "Little or not at all responsible" have been collapsed into one due to the lack of subjects in single categories or tertiles. EA: environmental association; SPGs: solidarity purchasing groups.

4. Discussion

This cross-sectional study gives information on the adherence to the MD in a representative sample of adults residing in Italy. Furthermore, this investigation sheds light on the subjects' evaluation of the Mediterranean dietary model as a sustainable diet based on the FAO's statement [33] and on whether the subjects perceived their dietary habits of the last 3 months to be sustainable. These two variables have not been explored in relation to the Mediterranean dietary model before, opening the way to further investigations on the relationship between the adoption of a Mediterranean dietary pattern and its subjective interpretation. In fact, to the best of our knowledge, this is the first time in which an association between the adherence to the MD and its evaluation as a sustainable dietary model, as well as between the adherence to the MD and the self-perception of dietary habits as sustainable, has been assessed. Indeed, exploring both consumers' diets and their perceptions about food consumption is a valuable approach to define effective strategies to opportunely shift dietary behavior in a desired direction. Specifically, the mechanism able to explain the association between self-perceptions of the MD's sustainability and the adoption of a sustainable diet with the adherence to the MD can reasonably rely on the consumers' awareness of their own dietary behaviors, their knowledge of food and environmental sustainability issues—concepts largely misunderstood by the general population [36]—and on the MD's relevance in terms of sustainability. Nevertheless, further research on the potential mechanisms involved in this association are needed. In our study, the MD has been scarcely recognized as a sustainable diet. Thus, given the growing focus and interest on the need to adopt sustainable behaviors and a positive attitude toward food sustainability [36,37], not considering the MD as a sustainable diet could limit the adherence to the MD itself.

In general, a medium adherence to the MD was reported, in line with another recent investigation on Italian adult populations [38]. Similar to what was found by Dinu et al. [38], better adherence to the MD was found in females compared with males. In addition, our results confirm the influence of the educational level, as less educated people showed lower adherence. However, contrary to what was shown by Dinu et al. [38] and by Caparello et al. [39], no associations were observed for age, nor between the compliance to the Mediterranean dietary model and the geographical area of residence. Such discrepancies with our results may reflect the lack of subjects' representativeness based on the age and geographical distribution of the participants in both of these studies. Our findings reflect the effects of globalization, which has shaped the dietary habits of people living in the Mediterranean Basin toward food consumption that traditionally has characterized non-Mediterranean countries [40]. However, conflicting data on trends of adherence to the MD have been reported. Indeed, while a cross-sectional investigation in South Italy, one of the MD cradles, found a significant decrease in adherence to the MD from the 1980s to the 2000s, mainly in younger groups [41], a study carried out with an adult population living in the north of Italy did not report a significant change in MD adherence from 1991 to 2006 [42].

Socioeconomic status emerged as a factor impacting participants' dietary habits. Indeed, people declaring higher incomes and education showed better compliance to the Mediterranean dietary pattern, contrary to those in the opposite conditions. Moreover, having a higher income clearly increased the probability for respondents obtaining a higher score of adherence to the MD. This data are confirmed by a recent previous work, which found that a less advantageous socioeconomic status represents an obstacle to following the MD [43]. Although it has been argued that, in principle, the weekly costs of the MD and Italian household consumption do not differ significantly [24], other findings suggest that greater adherence to the MD increases the monetary diet costs compared with less adherence. In Italy, the Moli-sani study pointed out the role of economic constraints in determining the low adherence to the MD in a period of economic crisis (2007–2010), with greater detrimental effects in the elderly [44]. On the other hand, the role of education may be explained by the influence of nutritional knowledge and awareness about the role of diet in promoting healthy lifestyles, as higher education is generally linked to healthy food consumption [45]. Similar to our results, a recent work carried out on Dutch adults found that highly educated individuals followed better consumption patterns compared with less educated ones [46].

According to the consumption frequency of single food groups, the majority of the respondents were not compliant with national [32] and international [47] dietary recommendations. Indeed, fruit and vegetable intakes were far below the suggested cut-offs (i.e., $\geq$400 g or $\geq$5 portions per day). Considering protein-based foods of animal origin, only approximately 30% of the respondents and about 2% among the subjects falling in the low adherence to the MD group declared eating fish or sea food as suggested both by the MD [31] and by the Italian nutritional guidelines [32] (i.e., $\geq$2 servings per week). Furthermore, the consumption of red meat exceeded 3 portions per week (the cut-off proposed by the Mediterranean dietary pattern) in a low proportion of subjects (<10%). Nevertheless, a higher number of subjects was not compliant with the stricter Italian guidelines, recommending one serving per week of fresh red meat and the occasional consumption of processed or cured meat. The wider inclusion of instances related to environmental sustainability in the national recommendations, being more plant-oriented, explains the discrepancy between the two cut-offs. Our study suggests that among the food components, the lowest level of compliance to the MD-based cut-offs can be observed for olive oil followed by legumes and fish not only in the whole sample of respondents, but also in the highest MD category. This data are in line with previous works, which reported a decrease in olive oil consumption over time [41], probably due to its limited affordability [48].

This study suggests that some changes are needed in the dietary behaviors of Italian adults in order to meet nutritional and environmental guidelines, as is also expressed by the adherence to the MD. From this perspective, public interventions might be defined, for instance, to increase consumers' awareness of the environmental impact of food choices [49], to nudge the consumer away from bad choices [50], or to modify the relative price of healthy or unhealthy choices with fiscal measures [51,52]. However, as has been suggested by several authors, the public decision-making process might be challenging given the uncertainties regarding outcomes due to the non-linear processes, feedback loops, and trade-offs that occur in food systems [53]. For instance, it has been shown that lowering domestic demand at the European level for meat would affect the profitability of meat production in the EU, in particular the European beef meat sector [54]. The complex network of interdependencies among food systems and dietary behavior outcomes asks for a multidisciplinary approach, requiring concerted efforts between disciplines.

To the best of our knowledge, this is the first study evaluating associations between MD adherence, subjects' awareness about MD sustainability, and self-perception about their own dietary behavior. Other strengths of this work should be highlighted. First of all, the study population is representative of the Italian adult population in terms of gender, age, and geographical distribution. In addition, important information that may impact food habits was collected (e.g., respondents' BMIs, nationalities, education levels, incomes, occupations, household sizes, degrees of responsibility for food purchasing and meal preparation, habitual frequency of eating out, and any participation in solidarity purchasing groups or environmental associations). Nevertheless, other potential determinants of dietary habits (e.g., physical activity level and smoking habits) were not investigated. Another limitation of this study is linked to the use of a self-administered online questionnaire, which is a very useful and economical tool but, at the same time, may lead to recall bias and misclassification. In addition, some questions (e.g., the self-perception of the MD's sustainability and the adoption of a sustainable diet) have not specifically been tested for validity, and this could represent a limitation for the soundness of the results in the present study. On the other hand, the use of a validated questionnaire specifically designed to collect information on adherence to the MD (main outcome) has been used. Furthermore, the distribution of the sample according to certain categorical variables in some case led to a very low number of subjects (e.g., respondents classified as having a little or lack of responsibility in purchasing or preparing food), limiting the analysis reliability.

5. Conclusions

The present research appraised MD adherence in a representative sample of adults residing in Italy, investigated the self-perception of adopting a sustainable diet, and expressed subjects' levels of agreement for considering the MD as a sustainable dietary model. Associations with anthropometric and sociodemographic variables revealed a series of contributing factors to MD adherence, primarily BMI, education attainment, and income level. Overall, the sample reported medium adherence to the MD, which is rooted in Italian culture and a widely recognized model of dietary sustainability. The results suggest a gradual shift away from this dietary pattern and support the need to address efforts for driving dietary transition and developing intervention strategies tailored to a target population of adults. There is a need to spark renewed interest in the general population in rediscovering a traditional dietary model that can boast nutritional, environmental, and social sustainability dimensions. Ensuring economic accessibility to food supplies should be the first public health prevention strategy for improving diet quality and increasing adherence to the MD. Furthermore, public campaigns should stress the link between diet and its environmental impact to foster nutritionally adequate and eco-friendly dietary behaviors.

Supplementary Materials: The following are available online at https://www.mdpi.com/article/10.3390/nu13093282/s1, Table S1: Percentage distribution of respondents according to their consumption frequency of single food groups/items.

Author Contributions: Conceptualization, B.B., A.R., D.M. and F.S.; methodology, B.B., A.R., D.M. and F.S.; software, B.B., A.R., and D.M.; formal Analysis, B.B., A.R. and D.M.; investigation, B.B., A.R., D.M. and F.S.; resources, F.S. and D.M.; data curation, B.B., A.R., D.M. and F.S.; writing—original draft preparation, B.B.; writing—review and editing, B.B., A.R., D.M. and F.S.; visualization, B.B., A.R., D.M. and F.S.; supervision, F.S.; project administration, F.S. and D.M. All authors have read and agreed to the published version of the manuscript.

Funding: This research received no external funding.

Institutional Review Board Statement: This study was conducted according to the guidelines of the Declaration of Helsinki and approved by the local institutional ethical committee (Comitato Etico Area Vasta Emilia Nord, approval code 1139/2018/OSS/UNIPR).

Informed Consent Statement: Informed consent was obtained from all subjects involved in the study.

Data Availability Statement: The data presented in this study are available on request from the corresponding author.

Acknowledgments: The authors thank Cristian Ricci for his statistical suggestions and Cinzia Franchini for her contribution to the graphical abstract.

Conflicts of Interest: The authors declare no conflict of interest.

References

1. Galbete, C.; Schwingshackl, L.; Schwedhelm, C.; Boeing, H.; Schulze, M.B. Evaluating Mediterranean diet and risk of chronic disease in cohort studies: An umbrella review of meta-analyses. *Eur. J. Epidemiol.* **2018**, *33*, 909–931. [CrossRef]
2. Sofi, F.; Abbate, R.; Gensini, G.F.; Casini, A. Accruing evidence on benefits of adherence to the Mediterranean diet on health: An updated systematic review and meta-analysis. *Am. J. Clin. Nutr.* **2010**, *92*, 1189–1196. [CrossRef]
3. Dinu, M.; Pagliai, G.; Angelino, D.; Rosi, A.; Dall'Asta, M.; Bresciani, L.; Ferraris, C.; Guglielmetti, M.; Godos, J.; Del Bo', C.; et al. Effects of Popular Diets on Anthropometric and Cardiometabolic Parameters: An Umbrella Review of Meta-Analyses of Randomized Controlled Trials. *Adv. Nutr.* **2020**, *11*, 815–833. [CrossRef] [PubMed]
4. Dinu, M.; Pagliai, G.; Casini, A.; Sofi, F. Mediterranean diet and multiple health outcomes: An umbrella review of meta-analyses of observational studies and randomised trials. *Eur. J. Clin. Nutr.* **2017**, *72*, 30–43. [CrossRef] [PubMed]
5. Georgoulis, M.; Kontogianni, M.D.; Yiannakouris, N. Mediterranean Diet and Diabetes: Prevention and Treatment. *Nutrients* **2014**, *6*, 1406–1423. [CrossRef] [PubMed]
6. Becerra-Tomás, N.; Mejía, S.B.; Viguiliouk, E.; Khan, T.; Kendall, C.W.; Kahleova, H.; Rahelić, D.; Sievenpiper, J.L.; Salas-Salvadó, J. Mediterranean diet, cardiovascular disease and mortality in diabetes: A systematic review and meta-analysis of prospective cohort studies and randomized clinical trials. *Crit. Rev. Food Sci. Nutr.* **2019**, *60*, 1207–1227. [CrossRef] [PubMed]
7. Godos, J.; Zappalà, G.; Bernardini, S.; Giambini, I.; Bes-Rastrollo, M.; Martinez-Gonzalez, M.A. Adherence to the Mediterranean diet is inversely associated with metabolic syndrome occurrence: A meta-analysis of observational studies. *Int. J. Food Sci. Nutr.* **2016**, *68*, 138–148. [CrossRef] [PubMed]
8. Schwingshackl, L.; Schwedhelm, C.; Galbete, C.; Hoffmann, G. Adherence to Mediterranean Diet and Risk of Cancer: An Updated Systematic Review and Meta-Analysis. *Nutrients* **2017**, *9*, 1063. [CrossRef]
9. Psaltopoulou, T.; Sergentanis, T.N.; Panagiotakos, D.B.; Sergentanis, I.N.; Kosti, R.; Scarmeas, N. Mediterranean diet, stroke, cognitive impairment, and depression: A meta-analysis. *Ann. Neurol.* **2013**, *74*, 580–591. [CrossRef]
10. Godos, J.; Ferri, R.; Caraci, F.; Cosentino, F.I.I.; Castellano, S.; Galvano, F.; Grosso, G. Adherence to the Mediterranean Diet is Associated with Better Sleep Quality in Italian Adults. *Nutrients* **2019**, *11*, 976. [CrossRef]
11. Willett, W.; Rockström, J.; Loken, B.; Springmann, M.; Lang, T.; Vermeulen, S.; Garnett, T.; Tilman, D.; DeClerck, F.; Wood, A.; et al. Food in the Anthropocene: The EAT–Lancet Commission on healthy diets from sustainable food systems. *Lancet* **2019**, *393*, 447–492. [CrossRef]
12. HLPE. *Nutrition and Food Systems. A Report by the High Level Panel Experts on Food Security Nutrition*; Committee on World Food Security: Rome, Italy, 2017; p. 152.
13. FAO; OIE; WHO; UN System Influenza Coordination; UNICEF; World Bank. *Contributing to One World, One Health: A Strategic Framework for Reducing Risks of Infectious Diseases at the Animal-Human-Ecosystems Interface, Consultation Document*; Food and Agriculture Organization; World Health Organization; UN System Influenza Coordinator United Nations Children Fund; World Bank: Sharm-el-Sheikh, Egypt, 2008.
14. Bach-Faig, A.; Berry, E.M.; Lairon, D.; Reguant, J.; Trichopoulou, A.; Dernini, S.; Medina, F.X.; Battino, M.; Belahsen, R.; Miranda, G.; et al. Mediterranean diet pyramid today. Science and cultural updates. *Public Health Nutr.* **2011**, *14*, 2274–2284. [CrossRef]

15. Dernini, S.; Berry, E.M. Mediterranean Diet: From a Healthy Diet to a Sustainable Dietary Pattern. *Front. Nutr.* **2015**, *2*, 15. [CrossRef]
16. Burlingame, B.; Dernini, S. Sustainable diets: The Mediterranean diet as an example. *Public Health Nutr.* **2011**, *14*, 2285–2287. [CrossRef]
17. Dernini, S.; Berry, E.; Serra-Majem, L.; La Vecchia, C.; Capone, R.; Medina, F.X.; Aranceta-Bartrina, J.; Belahsen, R.; Burlingame, B.; Calabrese, G.; et al. Med Diet 4.0: The Mediterranean diet with four sustainable benefits. *Public Health Nutr.* **2016**, *20*, 1322–1330. [CrossRef]
18. Tilman, D.; Clark, M. Global diets link environmental sustainability and human health. *Nature* **2014**, *515*, 518–522. [CrossRef] [PubMed]
19. Blas, A.; Garrido, A.; Unver, O.; Willaarts, B.A. A comparison of the Mediterranean diet and current food consumption patterns in Spain from a nutritional and water perspective. *Sci. Total. Environ.* **2019**, *664*, 1020–1029. [CrossRef] [PubMed]
20. UNESCO. Representative List of the Intangible Cultural Heritage of Humanity. Paris. 2010. Available online: http://www.unesco.org/culture/ich/RL/00884 (accessed on 30 June 2021).
21. Ruiz, A.H.; García-Villanova, B.; Hernández, E.J.G.; Amiano, P.; Azpiri, M.; Montes, E.M. Description of indexes based on the adherence to the Mediterranean dietary pattern: A review. *Nutr. Hosp.* **2015**, *32*, 1872–1884.
22. Bach, A.; Serra-Majem, L.; Carrasco, J.L.; Roman, B.; Ngo, J.; Bertomeu, I.; Obrador, B. The use of indexes evaluating the adherence to the Mediterranean diet in epidemiological studies: A review. *Public Health Nutr.* **2006**, *9*, 132–146. [CrossRef]
23. Rosi, A.; Paolella, G.; Biasini, B.; Scazzina, F.; Alicante, P.; De Blasio, F.; Russo, M.D.; Rendina, D.; Tabacchi, G.; Cairella, G.; et al. Dietary habits of adolescents living in North America, Europe or Oceania: A review on fruit, vegetable and legume consumption, sodium intake, and adherence to the Mediterranean Diet. *Nutr. Metab. Cardiovasc. Dis.* **2019**, *29*, 544–560. [CrossRef] [PubMed]
24. Germani, A.; Vitiello, V.; Giusti, A.M.; Pinto, A.; Donini, L.M.; Del Balzo, V. Environmental and economic sustainability of the Mediterranean Diet. *Int. J. Food Sci. Nutr.* **2014**, *65*, 1008–1012. [CrossRef]
25. Belahsen, R. Nutrition transition and food sustainability. *Proc. Nutr. Soc.* **2014**, *73*, 385–388. [CrossRef] [PubMed]
26. Benton, T.; Bieg, C.; Harwatt, H.; Pudasaini, R.; Wellesley, L. *Food System Impacts on Biodiversity Loss: Three Levers for Food System Transformation in Support of Nature*; Royal Institute of International Affairs: London, UK, 2021.
27. Biasini, B. Understanding, Assessing and Modelling Sustainable Eating Behaviours. Ph.D. Thesis, University of Parma, Parma, Italy, 2020.
28. WHO. Cut-Off for BMI According to WHO Standards 2010. Available online: https://gateway.euro.who.int/en/indicators/mn_survey_19-cut-off-for-bmi-according-to-who-standards/ (accessed on 30 June 2021).
29. Graziano, P.R.; Forno, F. Political Consumerism and New Forms of Political Participation. *Ann. Am. Acad. Politi.-Soc. Sci.* **2012**, *644*, 121–133. [CrossRef]
30. Vittersø, G.; Torjusen, H.; Laitala, K.; Tocco, B.; Biasini, B.; Csillag, P.; De Labarre, M.D.; Lecoeur, J.-L.; Maj, A.; Majewski, E.; et al. Short Food Supply Chains and Their Contributions to Sustainability: Participants' Views and Perceptions from 12 European Cases. *Sustainability* **2019**, *11*, 4800. [CrossRef]
31. Gnagnarella, P.; Dragà, D.; Misotti, A.; Sieri, S.; Spaggiari, L.; Cassano, E.; Baldini, F.; Soldati, L.; Maisonneuve, P. Validation of a short questionnaire to record adherence to the Mediterranean diet: An Italian experience. *Nutr. Metab. Cardiovasc. Dis.* **2018**, *28*, 1140–1147. [CrossRef] [PubMed]
32. CREA. Guidelines for Healthy Italian Nutrition. 2019. Available online: https://www.crea.gov.it/web/alimenti-e-nutrizione/-/linee-guida-per-una-sana-alimentazione-2018 (accessed on 30 June 2021).
33. Burlingame, B.; Dernini, S. Sustainable Diets and Biodiversity: Directions and Solutions for Policy, Research and Action. In Proceedings of the International Scientific Symposium, Biodiversity and Sustainable Diets United Against Hunger, Rome, Italy, 3–5 November 2010; Food and Agriculture Organization of the United Nations (FAO): Rome, Italy, 2012.
34. Riddell, L.J.; Ang, B.; Keast, R.S.; Hunter, W. Impact of living arrangements and nationality on food habits and nutrient intakes in young adults. *Appetite* **2011**, *56*, 726–731. [CrossRef] [PubMed]
35. Fishbein, M.; Ajzen, I. *Predicting and Changing Behavior: The Reasoned Action Approach*; Psychology Press: New York, NY, USA, 2010.
36. García-González, Á.; Achón, M.; Krug, A.C.; Varela-Moreiras, G.; Alonso-Aperte, E. Food Sustainability Knowledge and Attitudes in the Spanish Adult Population: A Cross-Sectional Study. *Nutrients* **2020**, *12*, 3154. [CrossRef]
37. Vermeir, I.; Weijters, B.; De Houwer, J.; Geuens, M.; Slabbinck, H.; Spruyt, A.; Verbeke, W. Environmentally sustainable food consumption: A review and research agenda from a goal-directed perspective. *Front. Psychol.* **2020**, *11*, 1603. [CrossRef] [PubMed]
38. Dinu, M.; Pagliai, G.; Giangrandi, I.; Colombini, B.; Toniolo, L.; Gensini, G.; Sofi, F. Adherence to the Mediterranean diet among Italian adults: Results from the web-based Medi-Lite questionnaire. *Int. J. Food Sci. Nutr.* **2020**, *72*, 271–279. [CrossRef]
39. Caparello, G.; Galluccio, A.; Giordano, C.; Lofaro, D.; Barone, I.; Morelli, C.; Sisci, D.; Catalano, S.; Andò, S.; Bonofiglio, D. Adherence to the Mediterranean diet pattern among university staff: A cross-sectional web-based epidemiological study in Southern Italy. *Int. J. Food Sci. Nutr.* **2019**, *71*, 581–592. [CrossRef] [PubMed]
40. Da Silva, R.; Bach-Faig, A.; Quintana, B.R.; Buckland, G.; De Almeida, M.D.V.; Serra-Majem, L. Worldwide variation of adherence to the Mediterranean diet, in 1961–1965 and 2000–2003. *Public Health Nutr.* **2009**, *12*, 1676–1684. [CrossRef]
41. Veronese, N.; Notarnicola, M.; Cisternino, A.M.; Inguaggiato, R.; Guerra, V.; Reddavide, R.; Donghia, R.; Rotolo, O.; Zinzi, I.; Leandro, G.; et al. Trends in adherence to the Mediterranean diet in South Italy: A cross sectional study. *Nutr. Metab. Cardiovasc. Dis.* **2019**, *30*, 410–417. [CrossRef] [PubMed]

42. Pelucchi, C.; Galeone, C.; Negri, E.; La Vecchia, C. Trends in adherence to the Mediterranean diet in an Italian population between 1991 and 2006. *Eur. J. Clin. Nutr.* **2010**, *64*, 1052–1056. [CrossRef]
43. Cavaliere, A.; Banterle, A.; De Marchi, E. Exploring the Adherence to the Mediterranean Diet and Its Relationship with Individual Lifestyle: The Role of Healthy Behaviors, Pro-Environmental Behaviors, Income, and Education. *Nutrients* **2018**, *10*, 141. [CrossRef] [PubMed]
44. Bonaccio, M.; Di Castelnuovo, A.; Bonanni, A.; Costanzo, S.; De Lucia, F.; Persichillo, M.; Zito, F.; Donati, M.B.; de Gaetano, G.; Iacoviello, L. Decline of the Mediterranean diet at a time of economic crisis. Results from the Moli-sani study. *Nutr. Metab. Cardiovasc. Dis.* **2014**, *24*, 853–860. [CrossRef] [PubMed]
45. Buscemi, S. What are the determinants of adherence to the mediterranean diet? *Int. J. Food Sci. Nutr.* **2021**, *72*, 143–144. [CrossRef]
46. Van Bussel, L.M.; Van Rossum, C.T.; Temme, E.H.; Boon, P.E.; Ocké, M.C. Educational differences in healthy, environmentally sustainable and safe food consumption among adults in the Netherlands. *Public Health Nutr.* **2020**, *23*, 2057–2067. [CrossRef] [PubMed]
47. WHO. *Promoting Fruit and Vegetable Consumption around the World*; Global Strategy on Diet, Physical Activity and Health; World Health Organization: Geneva, Switzerland, 2003.
48. Luchetti, F. Importance and future of olive oil in the world market—An introduction to olive oil. *Eur. J. Lipid Sci. Technol.* **2002**, *104*, 559–563. [CrossRef]
49. Abrahamse, W. How to Effectively Encourage Sustainable Food Choices: A Mini-Review of Available Evidence. *Front. Psychol.* **2020**, *11*, 589674. [CrossRef] [PubMed]
50. Sogari, G.; Li, J.; Lefebvre, M.; Menozzi, D.; Pellegrini, N.; Cirelli, M.; Gómez, M.I.; Mora, C. The Influence of Health Messages in Nudging Consumption of Whole Grain Pasta. *Nutrients* **2019**, *11*, 2993. [CrossRef]
51. Castellari, E.; Marette, S.; Moro, D.; Sckokai, P. Can menu labeling affect away-from-home-dietary choices? *Bio-Based Appl. Econ.* **2018**, *7*, 249–263.
52. Afshin, A.; Penalvo, J.L.; Del Gobbo, L.; Silva, J.; Michaelson, M.; O'Flaherty, M.; Capewell, S.; Spiegelman, D.; Danaei, G.; Mozaffarian, D. The prospective impact of food pricing on improving dietary consumption: A systematic review and meta-analysis. *PLoS ONE* **2017**, *12*, e0172277. [CrossRef] [PubMed]
53. Fresco, L.O.; Geerling-Eiff, F.; Hoes, A.-C.; van Wassenaer, L.; Poppe, K.J.; Vorst, J.G.A.J.V.D. Sustainable food systems: Do agricultural economists have a role? *Eur. Rev. Agric. Econ.* **2021**, *48*, 694–718. [CrossRef]
54. Santini, F.; Ronzon, T.; Perez Dominguez, I.; Araujo Enciso, S.R.; Proietti, I. What if meat consumption would decrease more than expected in the high-income countries? *Bio-Based Appl. Econ. J.* **2017**, *6*, 37–56.

Article

Adherence to the Mediterranean Diet in a Portuguese Immigrant Community in the Central Valley of California

Roberto M. Couto, Andrew D. Frugé and Michael W. Greene *

Department of Nutrition, Dietetics, and Hospitality Management, Auburn University, Auburn, AL 36849, USA; rmc0060@auburn.edu (R.M.C.); adf0003@auburn.edu (A.D.F.)
* Correspondence: mwgreene@auburn.edu; Tel.: +1-334-844-8435

Abstract: The Mediterranean Diet (MedDiet) is a healthy eating pattern associated with a better quality of life among older adults and reduced risk of non-communicable diseases. Little is known about the MedDiet in immigrant communities from countries in which the MedDiet is a settled cultural heritage. Thus, we examined MedDiet adherence and perceived knowledge, benefits, and barriers to the MedDiet in a Portuguese immigrant community in Turlock, California. A cross-sectional study was conducted with 208 participants in Turlock and Livermore, California, which was used as a reference population. Univariate, multivariable, and logistic regression models were used for data analysis. Compared to the Livermore group, the Turlock group was younger and less educated, but had a higher average MedDiet score and active adherence to a MedDiet ($p < 0.001$ for both). In the Turlock group, convenience, sensory appeal, and health were observed to be significant barriers to the MedDiet ($p < 0.05$), while health, weight loss, natural content, familiarity, price, sensory appeal, and mood were significant benefit factors ($p < 0.05$). In conclusion, participants in Turlock had greater MedDiet adherence despite lower education attainment. Furthermore, the perceived benefits of the MedDiet were key factors in MedDiet perception and adherence in a Portuguese immigrant community.

Keywords: Mediterranean diet; adherence; Portuguese immigrants; California

Citation: Couto, R.M.; Frugé, A.D.; Greene, M.W. Adherence to the Mediterranean Diet in a Portuguese Immigrant Community in the Central Valley of California. *Nutrients* **2021**, *13*, 1989. https://doi.org/10.3390/nu13061989

Academic Editor: Daniela Bonofiglio

Received: 29 April 2021
Accepted: 4 June 2021
Published: 9 June 2021

Publisher's Note: MDPI stays neutral with regard to jurisdictional claims in published maps and institutional affiliations.

1. Introduction

Nutrition is thought to be a core element in ameliorating health conditions in a globally aging society [1]. The Mediterranean Diet (MedDiet) has been widely reported as a model of a healthy eating pattern for a better quality of life among older adults [2,3] and for reducing the risk of the most prevalent non-communicable diseases associated with aging [4], such as cardiovascular disease [5–7], cancer [8], metabolic syndrome [6], obesity [5,6], and dementia [3]. The MedDiet is not universally defined because it is a dietary pattern from a relatively heterogeneous group of countries bordering the Mediterranean Sea and also within the countries themselves [6]. However, the traditional MedDiet is a plant-based diet that is characterized by high consumption of fruits, non-refined cereals, vegetables, olive oil, and nuts; a moderate amount of chicken and fish, and lower consumption of dairy, red meat, and sugars; use of aromatic herbs; water as the primary beverage; and wine in moderation [9].

Portugal is the most western country in southern Europe and is geographically not in the Mediterranean basin. Still, the Mediterranean diet is a settled cultural heritage of the Portuguese population and cuisine [10], with ancestral food-related influence from their Mediterranean neighbors and specificities from surrounding migrants [11,12]. MedDiet adherence in Portugal was similar to that of Spain for the period of 2004–2011 when assessed from Food and Agriculture Organization Food Balance Sheets [13]. Like Italy, Greece, and Spain, MedDiet adherence in Portugal was greatly reduced from 1961 to 1965 compared to 2004–2011 [13]. As observed in neighboring countries, regional differences in adherence to the Mediterranean food pattern exist in Portugal. The regions of Algarve and Madeira have the highest MedDiet adherence, while Lisbon and Tejo valley and the Azores have the lowest [14].

Portuguese immigration to the United States occurred in two major waves: the first from 1890 to 1930 and the second from 1950 to 1990 [15]. Approximately 1.37 million Portuguese Americans now live in the United States [16]. The majority of Portuguese immigrants were from the Azores [17] and settled primarily in Massachusetts and California, mainly in the Central Valley [15,17,18]. The Azores region of Portugal has one of the least educated populations in Portugal and is one of the poorest regions in Portugal [19,20]. Acculturation of Portuguese immigrants has been shown to be similar to that of Greek immigrants, yet Portuguese immigrants have historically had lower educational attainment and a slightly higher poverty rate [15]. Assessing this population's adherence to the MedDiet and the factors influencing these behaviors can provide meaningful insight from a public health perspective.

The application of behavior change models to diet can assist in determining appropriate measures for nutrition assessment, intervention, and outcome evaluation as well as the adoption of healthy behaviors and diets [21]. The transtheoretical model of behavior change focuses on behavior change being a dynamic process occurring in the following stages and processes of change (unaware, unengaged, deciding, decided no, decided yes, action, and maintenance) [22]. An individual's stage of change is influenced by beliefs, experiences, prior knowledge, and perceived benefits and barriers towards their behavioral change [23,24]. Perceived benefits and barriers towards adopting a dietary approach can be strong predictors of an individual's food choice and the individual's willingness to alter or adjust their current lifestyle [25].

Even though MedDiet adherence has previously been assessed in populations living throughout the world [2], including elderly Portuguese [26] and the United States [27–31], few studies have examined MedDiet adherence in communities with immigrants from countries in which the MedDiet is a settled cultural heritage. The current study aimed to assess MedDiet adherence and its associated factors in a Portuguese immigrant community in the Central Valley of California, USA.

2. Materials and Methods

2.1. Study Setting

The present study was conducted at two locations in Central California. Livermore, with a population of 90,189 [17] located in Alameda County, is the easternmost city in the San Francisco Bay Area and is considered a gateway community to the Central Valley. Livermore is a science and technology center that contains the Lawrence Livermore National Laboratory and Sandia National Laboratories and has a median household income of USD 116,942 [32]. In Livermore, 77.2% of the population is white, 1.8% African American, 11.1% Asian, and 20.2% Hispanic or Latino, 0.2% American Indian and Alaska Native [17]; 51.1% of the population is female [17]; 12.9% of the population is 65 years old and over, while 23.5% are under the age of 18; and 41.8% of the population holds a bachelor's degree or higher [17]. According to the United States Census American Community Survey, the top three reported ancestries reported in Livermore city are German, Irish, and English [33].

With a population of 73,631, Turlock is the second largest town in Stanislaus County, located in the rural San Joaquin Valley region of the Central Valley [32]. Turlock has a median household income of USD 56,639 [32]. Turlock contains one public university, California State University Stanislaus [18]. In Turlock, 76.8% of the population is white, 2.4% African American, 5.6% Asian, 37.1% Hispanic or Latino, 0.7% American Indian and Alaska Native [18]; 51.9% of the population is female; 13.0% of the population is 65 years old and over, while 26.8% are under the age of 18; and 24.5% of the population holds a bachelor's degree or higher and 81.1% hold a high school education or higher [18]. According to the United States Census American Community Survey, the top three reported ancestries reported in Turlock city are German, Portuguese, and English [33].

2.2. Survey Instrument

A previously validated survey instrument was used to assess MedDiet adherence, participants' stage of change, barriers, and benefits towards adopting the MedDiet and demographic variables [29]. Participants' MedDiet adherence was assessed using a validated [34] 14-question Mediterranean Diet Adherence Screener (MEDAS) that has been used to evaluate MedDiet adherence in Europe, including Portugal, and throughout the world [27,35–37]. Additionally, three questions were asked to assess participants' readiness to adopt a MedDiet using the Precaution Adoption Process Model (stages of change). Perceived barriers (18 questions; knowledge, convenience, sensory appeal, and health) and benefits (26 questions; knowledge, weight loss, ethical concerns, natural content, familiarity, price, sensory appeal, and mood) to the MedDiet were measured using a five-point Likert scale. Seven demographic and anthropometric questions determining the age, sex, ethnicity, height, weight, level of education, and previous nutrition education or knowledge were assessed. Body Mass Index (BMI) was calculated by dividing weight in kilograms (kg) by height in meter (m) squared. All survey questions were self-reported. The survey instrument was translated from English to Portuguese by a Portuguese language teacher and was back-translated from Portuguese to English by three native Portuguese speakers.

2.3. Survey Distribution

This study was approved by the Auburn University Institutional Review Board prior to distributing the surveys, which included language for inferred consent. Convenience sampling was used to obtain completed surveys from shoppers of Save Mart Supermarket stores in both Livermore and Turlock, California, from 14 October 2019 to 1 January 2020. Both Livermore and Turlock Save Mart Supermarket stores were open seven days a week, but only allowed sampling and data collection on Mondays and Saturdays. All Save Mart Supermarket Corporation rules and regulations for outside vendors were followed. Participants were not compensated for completing the survey instrument. All adults at least 45 years old in Livermore and Turlock were eligible for the study. Participants were provided the option of completing the survey in Portuguese. Approximately 37% of the participants residing in Turlock completed the survey in Portuguese. Two hundred and eleven participants completed surveys. Three surveys were excluded because participants did not answer all questions or did not meet the age requirement to participate in the study. The remaining 208 responses were a priori divided into two groups based on the geographical location of survey collection in California: Turlock (n = 125) and Livermore (n = 83).

2.4. Statistical Analyses

All data analyses were conducted with RStudio and the Rx64 3.6.0 software environment (RStudio, PBC, Boston, MA, USA). A crude (unadjusted) and multivariable backward stepwise linear regression analysis was used to assess the differences in total MedDiet adherence scores between the groups. Regression coefficient p values and main effect p values calculated using a type III Sum of Squares test were reported. A multivariate linear model was used to determine barriers and benefit question scores in the groups: the crude model was unadjusted and the adjusted model included all demographic variables. A backward stepwise elimination logistic regression was performed to identify the predictors of the stage of change with the demographic variables. Inclusion and retention criteria in the logistic regression model were set at p-value cutoff points of 0.25 and 0.10, respectively. A significance level of 0.05 was established.

3. Results

3.1. Demographic Assessment

We examined whether there were demographic differences between participants in the Turlock and Livermore groups. As shown in Table 1, significant differences ($p < 0.05$) in age, ethnicity, and education were observed between the participants in Turlock compared to Livermore. The Turlock group had a lower percentage of older adults (>65 years old), a greater percentage of participants in the 'other ethnic group' category, and a lower percentage of participants with associate's, bachelor's, and master's/profession degrees. There were no statistical differences between groups for sex, BMI, and health-related qualifications.

Table 1. Demographics of participants in Turlock and Livermore.

	Turlock † (*n* = 125)		**Livermore** † (*n* = 83)		
	n	%	*n*	%	*p*-Value
Sex *					0.724
Male	51	41	31	38	
Female	74	59	52	62	
Age *					**0.004**
45–54	17	13	5	6	
55–64	54	43	21	26	
65–74	39	31	34	41	
>75	16	12	22	27	
Ethnicity *					**0.010**
White	104	83	60	74	
Black	5	4	4	5	
Chinese	0	0	9	9	
Asian-other	0	0	1	1	
Other ethnic group	17	13	9	11	
Education *					**<0.001**
High School or lower	61	61.8	27	26.8	
GED	10	11.4	4	3.7	
Technical or trade certificate	6	5.7	9	8.5	
Associate's degree	6	5.7	24	24.4	
Bachelor's degree	8	7.3	27	26.8	
Master's or professional degree	10	8.1	9	9.8	
BMI *					0.166
Underweight	1	0	2	2.4	
Normal weight	39	31.7	27	40.2	
Overweight	57	43.1	26	24.4	
Obese	29	25.2	27	32.9	
Qualification *					0.321
Health or nutrition related qualifications	3	1.6	5	6.1	
No health or nutrition related qualifications	123	98.4	77	93.9	

* Significance across score categories by Pearson's chi-squared test; bold font indicates $p < 0.05$; † Turlock, California and Livermore, California. Demographic categories are indicated using grey background.

3.2. Mediterranean Diet Adherence

The total (MEDAS) score was analyzed using a crude and multivariable backward stepwise linear regression model adjusting for the demographic variables of sex and age. We observed an increase in the MEDAS scores in the Turlock group in comparison to the Livermore group in both the crude and adjusted models (Table 2). For each point increase in the MEDAS score in the Livermore group, an increase of 0.85 +/− 0.26 points ($p = 0.001$) was observed in the Turlock group in the crude model and 0.81 +/− 0.30 points ($p = 0.002$) in the adjusted model. When evaluating both demographic variables and MEDAS score, the MEDAS score was 0.55 +/− 0.25 points lower in males than females and 5.48 +/− 1.79 points higher in respondents older than 75. The demographic variables of education, BMI, or nutrition qualifications were not significant and did not improve the parsimoniousness of the lineal model.

Table 2. Multivariable linear regression analysis assessing Mediterranean diet adherence between groups adjusted for demographic categories, stages of change, barriers, and benefits. Linear regression analysis using a crude and multivariable backward stepwise model to assess Mediterranean diet adherence in the Turlock group.

		β	SE	*p*-Value *	Main Effects *p*-Value ‡
Crude Model					
Group	Livermore	Ref †			
	Turlock	0.85	0.26	**0.001**	
Backward Stepwise Model					
Group					**0.002**
	Livermore	Ref †			
	Turlock	0.81	0.30	**0.002**	
Sex					**0.030**
	Female	Ref †			
	Male	−0.55	0.25	**0.030**	
Age					**0.043**
	45–54	Ref †			
	55–64	0.14	0.34	0.737	
	65–74	−0.05	0.43	0.901	
	>75	5.48	1.79	**0.002**	

† Ref, reference group; * regression coefficient *p* value; ‡ Main effects were assessed by ANOVA using a type III Sum of Squares method; *p* values < 0.05 are indicated in bold font. Variable categories are indicated using grey background.

3.3. Barriers to Consuming a MedDiet in Turlock

Cronbach's alpha was calculated to assess the internal consistency of 18 questions sorted into four factors: knowledge, convenience, sensory appeal, and health. As shown in Table 3, The Knowledge barrier had a Cronbach's alpha = 0.43, which is below the 0.60 to 0.70 that is considered acceptable or adequate for assessing the internal consistency [38]. We did not remove questions to improve the reliability of the Knowledge barrier. The Convenience (Cronbach's alpha = 0.76), Sensory Appeal (Cronbach's alpha = 0.75), and Health barriers (Cronbach's alpha = 0.87) were internally valid. We used both an unadjusted (crude) and an adjusted linear regression model for sex, age, ethnicity, education, and BMI to assess knowledge, convenience, sensory appeal, and health barriers in the Turlock group compared to the Livermore group. Convenience ($\beta = 1.12$, SE = 0.50, $p = 0.027$), Sensory Appeal ($\beta = 0.69$, SE = 0.33, $p = 0.041$), and Health ($\beta = 0.39$, SE = 0.13, $p = 0.003$) were observed to be significant barriers to the MedDiet in the Turlock group (Table 3).

3.4. Benefits to Consuming a MedDiet in Turlock

Health, Weight Loss, Ethical Concerns, Natural Content, Familiarity, Price, Sensory Appeal, and Mood factors were used to assess the perceived benefits of adopting a MedDiet among survey participants. All eight barrier factors were internally valid (Health = 0.93; Weight Loss = 0.63; Ethical Concerns = 0.89; Natural Content 0.61; Familiarity = 0.80; Price = 0.91; Sensory Appeal = 0.79; Mood = 0.93). Similar to the assessment of barriers, we used both unadjusted and adjusted linear regression models for sex, age, ethnicity, education, and BMI to assess the benefits from adopting a MedDiet in the Turlock group using the Livermore group as a reference (Table 4). The Turlock group perceived the MedDiet to have more health benefits in the adjusted model (Health: $\beta = 3.72$, SE = 0.98, $p = <0.001$), and this association remained consistent in the crude model. Additionally, Familiarity (Familiarity: $\beta = 0.99$, SE 0.26, $p = <0.001$) was perceived to be a benefit in the Turlock group in both regression models. All benefit factors, except ethical concerns in the adjusted models, were significant ($p < 0.05$) across the three models in the Turlock group compared to the Livermore group.

Table 3. Crude and adjusted linear analysis of perceived MD barriers.

	Crude †			Adjusted ††		
Barrier	**β**	**SE**	***p*-Value**	**β**	**SE**	***p*-Value**
Knowledge ($n = 4$) ‡ (Cronbach's Alpha = 0.43)						
Livermore ▽	Ref			Ref		
Turlock	0.49	0.37	0.187	0.75	0.43	0.086
Convenience ($n = 4$) (Cronbach's Alpha = 0.76)						
Livermore	Ref			Ref		
Turlock	1.38	0.44	**0.001**	1.12	0.50	0.027
Sensory Appeal ($n = 3$) (Cronbach's Alpha = 0.75)						
Livermore	Ref			Ref		
Turlock	0.87	0.30	0.004	0.69	0.33	**0.041**
Health ($n = 4$) (Cronbach's Alpha = 0.87)						
Livermore	Ref			Ref		
Turlock	0.59	0.40	0.147	0.39	0.13	0.003

‡ Number of questions in each factor; * p values < 0.05 from type III Sum of Squares method are indicated in bold font; † Crude linear model; †† Adjusted linear model for sex, age, ethnicity, education, and BMI; ▽ Livermore was used as the reference (Ref) group in the linear model.

Table 4. Crude and adjusted linear analysis of perceived MD benefits.

	Crude †			Adjusted ††		
Benefits	**β**	**SE**	***p*-Value** *	**β**	**SE**	***p*-Value** *
Health ($n = 10$) ‡ (Cronbach's Alpha = 0.93)						
Livermore▽	Ref			Ref		
Turlock	3.89	0.84	**<0.001**	3.72	0.98	**<0.001**
Weight Loss ($n = 2$) (Cronbach's Alpha = 0.63)						
Livermore	Ref			Ref		
Turlock	0.78	0.19	**<0.001**	0.74	0.22	0.001
Ethical ($n = 2$) (Cronbach's Alpha = 0.89)						
Livermore	Ref			Ref		
Turlock	0.30	0.13	0.018	0.20	0.34	0.560
Natural Content ($n = 2$) (Cronbach's Alpha = 0.61)						
Livermore	Ref			Ref		
Turlock	0.57	0.18	0.001	0.66	0.20	**0.001**
Familiarity ($n = 2$) (Cronbach's Alpha= 0.80)						
Livermore	Ref			Ref		
Turlock	0.98	0.21	**<0.001**	0.99	0.26	**<0.001**
Price ($n = 2$) (Cronbach's Alpha = 0.91)						
Livermore	Ref			Ref		
Turlock	0.56	0.22	**0.013**	0.58	0.26	**0.037**
Sensory Appeal ($n = 2$) (Cronbach's Alpha = 0.79)						
Livermore	Ref			Ref		
Turlock	0.53	0.22	**0.017**	0.56	0.26	**0.041**
Mood ($n = 3$) (Cronbach's Alpha = 0.93)						
Livermore	Ref			Ref		
Turlock	1.11	0.34	0.002	1.01	0.42	**0.017**

‡ Number of questions in each factor; * p values < 0.05 from type III Sum of Squares method are indicated in bold font; † Crude linear model; †† Adjusted linear model for sex, age, ethnicity, education, and BMI; ▽ Livermore was used as the reference (Ref) group in the linear model.

3.5. Stages of Change and Demographic Influences

The distribution of participants across stages of change between the Turlock and Livermore groups was significantly different ($p < 0.001$) (Table 5). More participants in the Action/Maintenance category were observed in the Turlock group. In contrast, the Livermore group had more participants in the Unaware/Unengaged, Deciding, and Decided No category than the Turlock group ($p < 0.001$) (Table 5).

We performed logistic regression to examine the effect of demographic variables on the likelihood of being in each stage of change towards adopting the MedDiet in the whole cohort (Table 6). Participants were significantly less likely to be in the Unengaged/Unaware stage if they were from the Turlock cohort (OR = 0.23, 95% CI: 0.11–0.45, p = <0.001). Additionally, participants who had a master's or professional degree were shown to be less likely to be in the Unaware/Unengaged stage of change (OR = 0.20, 95% CI: 0.05–0.63, $p < 0.01$). Participants categorized as Black (OR = 9.27, 95% CI: 1.50–180, p = <0.05) were significantly more likely to be in the Unaware/Unengaged stage. In regard to the Deciding and Deciding Yes groups, no significant associations were observed. In contrast, participants with a GED education were more likely to be in the Deciding No stage of change (OR = 9.36, 95% CI: 1.13–65.3, p <0.05). Lastly, the Turlock cohort was significantly more likely to be in the Action/Maintenance group (OR = 17.7, 95% CI: 6.46–64.6). However, participants who are classified as having a master's or professional degree (OR = 5.59, 95% CI: 1.71–21.0) were more likely to be in the Action/Maintenance group.

Table 5. Percent of participants in the Livermore and Turlock groups by stage of change.

Stages of Change	Livermore	Turlock
Unaware/Unengaged	67.1	37.4
Deciding	17.0	4.9
Decided No	6.1	0.8
Decided Yes *	4.9	6.5
Action/Maintenance *	4.9	49.6

* Significance across score categories by Pearson's chi-squared test ($p < 0.05$).

Table 6. Backward stepwise elimination logistic regression of stage of change by demographic factors.

	Stages of Change				
	Unaware/Unengaged	Deciding	Decided Yes	Decided No	Action/Maintenance
	OR (95% CI)	OR (95% CI)	OR (95% CI)	OR (95% CI)	OR (95% CI)
Group					
Turlock	0.23 (0.11–0.45) ***			0.17 (0.20–0.88)	17.7 (6.46–64.6) ***
Sex					
Female	-	1.55 (0.45–0.99)	-	-	-
Age					
55–64	0.55 (0.25–1.20)	-	-	-	-
65–74	0.54 (0.25–1.18)	1.59 (0.68–3.64)	-	-	-
>75	-	-	-	-	-
Ethnicity					
Black	9.27 (1.50–180)*	-	-	-	-
Chinese	-	-	-	-	-
Asian	3.06 (0.58–23.7)	-	-	-	-
Other	-	-	-	-	-
Education					
GED	-	-	-	9.36 (1.13–65.3) *	-
Certificate	-	-	-	-	-
Associate's	-	0.40 (0.06–1.50)	-	-	-
Bachelor's	0.42 (0.16–1.04)	-	-	-	-
Master's or professional	0.20 (0.05–0.63) **	0.34 (0.02–1.80)	-	-	5.59 (1.71–21.0) **
BMI					
Underweight	-	-	-	-	-
Overweight	-	-	-	-	-
Obese	-	-	-	-	-

* p-value <0.05; ** p-value <0.01; *** p-value <0.001; - Not applicable.

4. Discussion

MedDiet adherence and related factors affecting adherence have not previously been studied among Portuguese immigrants in the United States. Therefore, we used a recently developed survey instrument to assess participants' MedDiet adherence, the participants' stage of change towards incorporating the MedDiet in their lifestyle, and perceived benefits and barriers to consuming a MedDiet [29]. To ensure the inclusion of Portuguese adult immigrants in the Turlock group, we surveyed adults older than 45 years old at the local Save Mart store in Turlock and provided the participants the opportunity to complete the survey in Portuguese. Indeed, 37% of the Turlock participants preferred to take the survey in Portuguese, while none of the participants identified as Portuguese in Livermore. The population in the present study had a greater percentage of female respondents than reported in the Livermore and Turlock communities [17]. In contrast, the Livermore population in the present study had comparable ethnicity and education demographics compared to those in the Livermore community [17]. In the Turlock population in the present study, comparable ethnicity but not education (less participants in the study with holding a bachelor's degree) to that in the Turlock community [17] was observed.

We found that the Turlock group had higher adherence to the MedDiet as assessed by the MEDAS score compared to the Livermore group. Similarly, a higher percentage of participants in the Turlock group were in the Action/Maintenance stage of change compared to the Livermore group. This result was surprising given that the majority of respondents in the Turlock group had a high school education or lower and that educational level has been shown to be negatively associated with MedDiet adherence in studies performed in the United States [29] and countries bordering the Mediterranean Sea [39,40]. In contrast, our results are consistent with a study in Portugal where Portuguese households from lower social classes had greater adherence to the Mediterranean food pattern compared to participants at higher social levels [14]. Our findings of higher MedDiet adherence scores in Turlock could be due to the "Healthy Immigrant Effect" where immigrants are, on average, healthier than native-born participants in a given population [41]. Whether diet knowledge, diet decision making, or the ability of immigrant families to overcome diet barriers are factors [42] contributing to greater adherence to the MedDiet in the Turlock group requires formal examination. Even though Turlock participants were significantly more likely to be in the Action/Maintenance stage of change, we did observe that approximately one-third of participants in Turlock were in the Unaware/Unengaged stage of change. Whether exposure to dietary acculturation, lack of interaction with family or support groups, length of years spent in the United States, immigration status (first vs. second-generation immigrant), or other social/environmental factors that influence acculturation [42,43] explain our findings with Turlock participants in the Unaware/Unengaged stage of change requires further investigation.

The perceived benefit of Familiarity and perceived barriers of Convenience and Sensory Appeal scores in the Turlock group were significantly higher than those in the Livermore group. The difference in Familiarity may be due to the Turlock participants' knowledge of the Portuguese cultural cuisine (high fish, fruit, and cheese consumption, moderate wine consumption, and low processed food consumption [44]) which are aligned with most of the dietary components of the MedDiet [45,46], while the differences in Convenience and Sensory Appeal scores in the Turlock group may be due to an inability to obtain culturally familiar and fresh food in an impoverished rural region of California. Our findings are consistent with a prior study in Israel examining Ethiopian immigrants' perceived benefits of diet and sensory responses while attempting to maintain current dietary patterns, which found that immigrants' choices are both guided through convenience and familiarity [47]. Sensory appeal and convenience have also been found to be significant factors related to the unwillingness of patients with nonalcoholic fatty liver disease in a Northern European community to adjust their diet patterns to a Mediterranean diet [48].

A surprising result in perceived benefits was that weight loss was seen as more of a benefit in the Turlock group rather than the Livermore group. This result is not consistent with prior findings that weight loss as a perceived benefit was observed in participants with low MedDiet adherence [29]. Whether our current findings on perceived weight loss are specific for immigrants from countries where the Mediterranean diet is a settled cultural heritage will require future studies. Our findings that participants with a Master's or Professional Degree were over five times more likely to be in the Action/Maintenance stage and had a lower OR of being in the Unaware/Unengaged stage and that participants were nine times more likely to be in the Decided No category if the participant had a GED are consistent with previous studies showing a strong correlation between education level and MedDiet adherence [27,29,49–51].

While we took a systematic approach using validated questionnaires to conduct this study, several limitations are acknowledged. We recruited a convenience sample from one Portuguese immigrant population and one reference city; thus, in addition to our relatively small sample size, additional studies with larger Portuguese immigrant populations in other geographic locations are needed to confirm our findings. Similarly, our sample was restricted to adults aged 50 years and older; therefore, the results cannot be applied to the general adult population. Another possible limitation is that the self-reported data in our study, which include weight, height, and dietary assessment, may not reflect actual values. Lastly, we did not formally test the association between immigrant status and high MedDiet adherence in immigrant groups in the United States.

5. Conclusions

The present study identifies a significant increase in MedDiet scores between participants in Turlock and participants in Livermore, which was consistent with a greater proportion of Turlock participants in the Action/Maintenance stage of change. Although high educational levels are strongly associated with higher adherence to the MedDiet, the Turlock group had lower education attainment than the Livermore group, yet had greater MedDiet adherence. Importantly, the benefits of health, weight loss, natural content, familiarity, price, sensory appeal, and mood are key factors in the perception of the MedDiet in the Turlock group. In contrast, convenience, sensory appeal, and health were observed to be significant barriers to the MedDiet in the Turlock group.

Author Contributions: Conceptualization, R.M.C. and M.W.G.; methodology, R.M.C., A.D.F., and M.W.G.; software, R.M.C. and M.W.G.; validation, R.M.C., A.D.F., and M.W.G.; formal analysis, R.M.C. and M.W.G.; investigation, R.M.C. and M.W.G.; data curation, R.M.C.; writing—original draft preparation, R.M.C. and M.W.G.; writing—review and editing, R.M.C., A.D.F., and M.W.G.; visualization, M.W.G.; supervision, M.W.G.; project administration, M.W.G.; funding acquisition, M.W.G. All authors have read and agreed to the published version of the manuscript.

Funding: This work was supported by the Alabama Agricultural Experiment Station and the Hatch program of the National Institute of Food and Agriculture, U.S. Department of Agriculture (Project number ALA044-1-18037) (M.W.G.).

Institutional Review Board Statement: The study was conducted according to the guidelines of the Declaration of Helsinki, and approved by the Institutional Review Board of Auburn University (protocol code # 19-350 EX 1908, approved on 26 August 2019).

Informed Consent Statement: Informed consent was obtained from all subjects involved in the study.

Data Availability Statement: The data presented in this study are available on request from the corresponding author.

Conflicts of Interest: The authors declare no conflict of interest. The funders had no role in the design of the study; in the collection, analyses, or interpretation of data; in the writing of the manuscript, or in the decision to publish the result.

References

1. Ruthsatz, M.; Candeias, V. Non-communicable disease prevention, nutrition and aging. *Acta Biomed.* **2020**, *91*, 379–388.
2. Dinu, M.; Pagliai, G.; Casini, A.; Sofi, F. Mediterranean diet and multiple health outcomes: An umbrella review of meta-analyses of observational studies and randomised trials. *Eur. J. Clin. Nutr.* **2018**, *72*, 30–43. [CrossRef] [PubMed]
3. Hernández-Galiot, A.; Goñi, I. Adherence to the Mediterranean diet pattern, cognitive status and depressive symptoms in an elderly non-institutionalized population. *Nutr. Hosp.* **2017**, *34*, 338–344. [CrossRef]
4. Kennedy, B.K.; Berger, S.L.; Brunet, A.; Campisi, J.; Cuervo, A.M.; Epel, E.S.; Franceschi, C.; Lithgow, G.J.; Morimoto, R.I.; Pessin, J.E.; et al. Geroscience: Linking aging to chronic disease. *Cell* **2014**, *159*, 709–713. [CrossRef] [PubMed]

5. Ahmad, S.; Moorthy, M.V.; Demler, O.V.; Hu, F.B.; Ridker, P.M.; Chasman, D.I.; Mora, S. Assessment of risk factors and biomarkers associated with risk of cardiovascular disease among women consuming a Mediterranean diet. *JAMA Netw. Open* **2018**, *1*, e185708. [CrossRef] [PubMed]
6. Martinez-Lacoba, R.; Pardo-Garcia, I.; Amo-Saus, E.; Escribano-Sotos, F. Mediterranean diet and health outcomes: A systematic meta-review. *Eur. J. Public Health* **2018**, *28*, 955–961. [CrossRef] [PubMed]
7. Sánchez-Taínta, A.; Estruch, R.; Bullo, M.; Corella, D.; Gomez-Gracia, E.; Fiol, M.; Algorta, J.; Covas, M.I.; Lapetra, J.; Zazpe, I.; et al. Adherence to a Mediterranean-type diet and reduced prevalence of clustered cardiovascular risk factors in a cohort of 3204 high-risk patients. *Eur. J. Cardiovasc. Prev. Rehabil.* **2008**, *15*, 589–593. [CrossRef]
8. Schwingshackl, L.; Schwedhelm, C.; Galbete, C.; Hoffmann, G. Adherence to Mediterranean diet and risk of cancer: An updated systematic review and meta-analysis. *Nutrients* **2017**, *9*, 1063. [CrossRef] [PubMed]
9. Davis, C.; Bryan, J.; Hodgson, J.; Murphy, K. Definition of the Mediterranean diet; a literature review. *Nutrients* **2015**, *7*, 9139–9153. [CrossRef]
10. Reguant-Aleix, J.; Sensat, F. The Mediterranean diet, intangible cultural heritage of humanity. In *Mediterra 2012*; CIHEAM–Sciences Po Les Presses: Paris, France, 2012; pp. 465–484.
11. Kittler, P.G.; Sucher, K.P.; Nelms, M. *Food and Culture*; Cengage Learning: Boston, MA, USA, 2011.
12. Vareiro, D.; Bach-Faig, A.; Quintana, B.R.; Bertomeu, I.; Buckland, G.; de Almeida, M.D.V.; Serra-Majem, L. Availability of Mediterranean and non-Mediterranean foods during the last four decades: Comparison of several geographical areas. *Public Health Nutr.* **2009**, *12*, 1667–1675. [CrossRef] [PubMed]
13. Vilarnau, C.; Stracker, D.M.; Funtikov, A.; da Silva, R.; Estruch, R.; Bach-Faig, A. Worldwide adherence to Mediterranean diet between 1960 and 2011. *Eur. J. Clin. Nutr.* **2019**, *72*, 83–91. [CrossRef]
14. Rodrigues, S.; Caraher, M.; Trichopoulou, A.; De Almeida, M. Portuguese households' diet quality (adherence to Mediterranean food pattern and compliance with WHO population dietary goals): Trends, regional disparities and socioeconomic determinants. *Eur. J. Clin. Nutr.* **2008**, *62*, 1263–1272. [CrossRef] [PubMed]
15. Scott, D.M. Portuguese Americans' acculturation, socioeconomic integration, and amalgamation: How far have they advanced? *Sociol. Probl. Prat.* **2009**, *61*, 41–64.
16. Rodrigues, I.; Pan, Y.; Lubkemann, S. observing census enumeration of non-english speaking households in 2010 Census: Portuguese report. *Surv. Methodol.* **2013**, *2013*, 15.
17. Pap, L. *The Portuguese-Americans*; Twayne Publishers: Boston, MA, USA, 1981.
18. Williams, J.R. And yet they come: Portuguese Immigration from the Azores to the United States: Center for Migration Studies of New York. *Can. J. Sociol. Cah. Can. Sociol.* **1982**, *9*, 215–217.
19. OECD Regional Economy [Internet]. 2015. Available online: https://www.oecd-ilibrary.org/content/data/6b288ab8-en (accessed on 30 November 2020).
20. OECD Regional Education [Internet]. 2019. Available online: https://www.oecd-ilibrary.org/content/data/213e806c-en (accessed on 30 November 2020).
21. Spahn, J.M.; Reeves, R.S.; Keim, K.S.; Laquatra, I.; Kellogg, M.; Jortberg, B.; Clark, N.A. State of the evidence regarding behavior change theories and strategies in nutrition counseling to facilitate health and food behavior change. *J. Am. Diet. Assoc.* **2010**, *110*, 879–891. [CrossRef]
22. Prochaska, J.O.; Johnson, S.; Lee, P. The transtheoretical model of behavior change. In *Handbook of Health Behavior Change*; Springer Publishing Company: Washington, DC, USA, 2009.
23. Mohr, P.; Quinn, S.; Morell, M.; Topping, D. Engagement with dietary fibre and receptiveness to resistant starch in Australia. *Public Health Nutr.* **2010**, *13*, 1915–1922. [CrossRef] [PubMed]
24. Weinstein, N.D.; Sandman, P.M. A model of the precaution adoption process: Evidence from home radon testing. *Health Psychol.* **1992**, *11*, 170. [CrossRef] [PubMed]
25. Pollard, J.; Kirk, S.L.; Cade, J.E. Factors affecting food choice in relation to fruit and vegetable intake: A review. *Nutr. Res. Rev.* **2002**, *15*, 373–387. [CrossRef] [PubMed]
26. Teixeira, B.; Afonso, C.; Sousa, A.S.; Guerra, R.S.; Santos, A.; Borges, N.; Moreira, P.; Padrão, P.; Amaral, T.F. Adherence to a Mediterranean dietary pattern status and associated factors among Portuguese older adults: Results from the nutrition UP 65 cross-sectional study. *Nutrition* **2019**, *65*, 91–96. [CrossRef]
27. Bottcher, M.R.; Marincic, P.Z.; Nahay, K.L.; Baerlocher, B.E.; Willis, A.W.; Park, J.; Gaillard, P.; Greene, M.W. Nutrition knowledge and Mediterranean diet adherence in the southeast United States: Validation of a field-based survey instrument. *Appetite* **2017**, *111*, 166–176. [CrossRef]
28. Chen, M.; Howard, V.; Harrington, K.F.; Creger, T.; Judd, S.E.; Fontaine, K.R. Does adherence to mediterranean diet mediate the association between food environment and obesity among non-hispanic black and white older US Adults? A path analysis. *Am. J. Health Promot.* **2020**, *34*, 652–658. [CrossRef] [PubMed]
29. Knight, C.J.; Jackson, O.; Rahman, I.; Burnett, D.O.; Frugé, A.D.; Greene, M.W. The Mediterranean diet in the Stroke Belt: A cross-sectional study on adherence and perceived knowledge, barriers, and benefits. *Nutrients* **2019**, *11*, 1847. [CrossRef] [PubMed]
30. Taylor, M.K.; Mahnken, J.D.; Sullivan, D.K. NHANES 2011–2014 reveals cognition of US older adults may benefit from better adaptation to the Mediterranean diet. *Nutrients* **2020**, *12*, 1929. [CrossRef]

31. Reedy, J.; Krebs-Smith, S.M.; Miller, P.E.; Liese, A.D.; Kahle, L.L.; Park, Y.; Subar, A.F. Higher diet quality is associated with decreased risk of all-cause, cardiovascular disease, and cancer mortality among older adults. *J. Nutr.* **2014**, *144*, 881–889. [CrossRef]
32. 2019 American Community Survey 5-Year Estimates [Internet]. 2019. Available online: https://data.census.gov/cedsci/profile?g=1600000US0680812 (accessed on 13 December 2020).
33. Selected Social Characteristics in the United States, 2011–2015 American Community Survey 5-Year Estimates [Internet]. 2015. Available online: https://www1.nyc.gov/assets/planning/download/pdf/planning-level/nyc-population/acs/soc_2015acs5yr_nyc.pdf. (accessed on 29 November 2020).
34. Schröder, H.; Fitó, M.; Estruch, R.; Martínez-González, M.A.; Corella, D.; Salas-Salvadó, J.; Lamuela-Raventós, R.; Ros, E.; Lamuela-Raventos, I.; Fiol, M.; et al. A short screener is valid for assessing Mediterranean diet adherence among older Spanish men and women. *J. Nutr.* **2011**, *141*, 1140–1145. [CrossRef]
35. García-Conesa, M.-T.; Philippou, E.; Pafilas, C.; Massaro, M.; Quarta, S.; Andrade, V.; Jorge, R.; Chervenkov, M.; Ivanova, T.; Dimitrova, D. Exploring the validity of the 14-item Mediterranean diet adherence screener (medas): A cross-national study in seven european countries around the Mediterranean region. *Nutrients* **2020**, *12*, 2960. [CrossRef]
36. Hebestreit, K.; Yahiaoui-Doktor, M.; Engel, C.; Vetter, W.; Siniatchkin, M.; Erickson, N.; Halle, M.; Kiechle, M.; Bischoff, S.C. Validation of the German version of the Mediterranean Diet Adherence Screener (MEDAS) questionnaire. *BMC Cancer* **2017**, *17*, 341. [CrossRef]
37. Papadaki, A.; Johnson, L.; Toumpakari, Z.; England, C.; Rai, M.; Toms, S.; Penfold, C.; Zazpe, I.; Martínez-González, M.A.; Feder, G. Validation of the English version of the 14-item Mediterranean Diet Adherence Screener of the PREDIMED study, in people at high cardiovascular risk in the UK. *Nutrients* **2018**, *10*, 138. [CrossRef] [PubMed]
38. Vaske, J.J.; Beaman, J.; Sponarski, C.C. Rethinking internal consistency in Cronbach's alpha. *Leis. Sci.* **2017**, *39*, 163–173. [CrossRef]
39. Cavaliere, A.; De Marchi, E.; Banterle, A. Exploring the adherence to the Mediterranean diet and its relationship with individual lifestyle: The role of healthy behaviors, pro-environmental behaviors, income, and education. *Nutrients* **2018**, *10*, 141. [CrossRef] [PubMed]
40. Ruggiero, E.; Di Castelnuovo, A.; Costanzo, S.; Persichillo, M.; Bracone, F.; Cerletti, C.; Donati, M.B.; de Gaetano, G.; Iacoviello, L.; Bonaccio, M. Socioeconomic and psychosocial determinants of adherence to the Mediterranean diet in a general adult Italian population. *Eur. J. Public Health* **2019**, *29*, 328–335. [CrossRef]
41. Markides, K.S.; Rote, S. The healthy immigrant effect and aging in the United States and other western countries. *Gerontologist* **2019**, *59*, 205–214. [CrossRef] [PubMed]
42. De Castro, J. Education of portuguese emigrants and their acculturation. *E-Methodology* **2017**, *4*, 55–74.
43. Castro, J.; Rudmin, F.W. Acculturation, acculturative change, and assimilation: A research bibliography with URL links. *Online Read. Psychol. Cult.* **2017**, *8*, 2307–0919.1075.
44. Robertson, C.; Robertson, D. Portuguese Cooking: The Authentic and Robust Cuisine of Portugal. In *Journal and Cookbook*; North Atlantic Books: Berleley, CA, USA, 1993.
45. Medina, F.X. Mediterranean diet, culture and heritage: Challenges for a new conception. *Public Health Nutr.* **2009**, *12*, 1618–1620. [CrossRef]
46. Real, H.; Dias, R.R.; Graça, P. Mediterranean diet conceptual model and future trends of its use in Portugal. *Health Promot. Int.* **2020**, *36*, 548–560. [CrossRef] [PubMed]
47. Leshem, M.; Dessie-Navon, H. Acculturation of immigrant diet, basic taste responses and sodium appetite. *J. Nutr. Sci.* **2018**, *7*, e21. [CrossRef] [PubMed]
48. Haigh, L.; Bremner, S.; Houghton, D.; Henderson, E.; Avery, L.; Hardy, T.; Hallsworth, K.; McPherson, S.; Anstee, Q.M. Barriers and facilitators to Mediterranean diet adoption by patients with nonalcoholic fatty liver disease in Northern Europe. *Clin. Gastroenterol. Hepatol.* **2019**, *17*, 1364–1371. [CrossRef] [PubMed]
49. Holgado, B.; de Irala-Estevez, J.; Martinez-Gonzalez, M.; Gibney, M.; Kearney, J.; Martinez, J. Barriers and benefits of a healthy diet in Spain: Comparison with other European member states. *Eur. J. Clin. Nutr.* **2000**, *54*, 453–459. [CrossRef]
50. Greiner, B.; Wheeler, D.; Croff, J.; Miller, B. Prior knowledge of the Mediterranean Diet is associated with dietary adherence in cardiac patients. *J. Am. Osteopath Assoc.* **2019**, *119*, 183–188. [CrossRef] [PubMed]
51. Bonaccio, M.; Bonanni, A.E.; Di Castelnuovo, A.; De Lucia, F.; Donati, M.B.; De Gaetano, G.; Iacoviello, L. Low income is associated with poor adherence to a Mediterranean diet and a higher prevalence of obesity: Cross-sectional results from the Moli-sani study. *BMJ Open* **2012**, *2*. [CrossRef] [PubMed]

MDPI

Article

Adherence to the Mediterranean Diet Is Associated with Better Metabolic Features in Youths with Type 1 Diabetes

Valentina Antoniotti [1], Daniele Spadaccini [2], Roberta Ricotti [1], Deborah Carrera [3], Silvia Savastio [1], Filipa Patricia Goncalves Correia [2], Marina Caputo [2], Erica Pozzi [1], Simonetta Bellone [1], Ivana Rabbone [1] and Flavia Prodam [2,4,*]

1 SCDU of Pediatrics, Department of Health Sciences, University of Piemonte Orientale, 28100 Novara, Italy; valentina.antoniotti@uniupo.it (V.A.); roberta.ricotti@uniupo.it (R.R.); savastio.silvia@gmail.com (S.S.); ericapozzi@yahoo.it (E.P.); simonetta.bellone@med.uniupo.it (S.B.); ivana.rabbone@uniupo.it (I.R.)

2 Department of Health Sciences, University of Piemonte Orientale, 28100 Novara, Italy; daniele.spadaccini@uniupo.it (D.S.); filipa.p.g.correia@gmail.com (F.P.G.C.); marina.caputo@uniupo.it (M.C.)

3 Pediatric Endocrinology, Azienda Ospedaliero Universitaria Maggiore Della Carità, 28100 Novara, Italy; carrera.deborah@gmail.com

4 Endocrinology, Department of Translational Medicine, University of Piemonte Orientale, 28100 Novara, Italy

* Correspondence: flavia.prodam@med.uniupo.it; Tel.: +39-0321-660693

Abstract: Our aim was to evaluate adherence to the Mediterranean diet (MedDiet) among children and adolescents with type 1 diabetes (T1D) in relation to metabolic control. Adherence to the MedDiet was assessed with the Mediterranean Diet Quality Index (KIDMED) questionnaire and physical activity by the International Physical Activity Questionnaire for Adolescent (IPAQ-A) on 65 subjects (32 males, 9–18 years) with T1D. Clinical and metabolic evaluation was performed (standardized body mass index (BMI-SDS), hemoglobin A1C (HbA1c), continuous glucose monitoring metrics when present, blood pressure, lipid profile). Parental characteristics (age, body mass index (BMI), socio-economic status) were reported. The adherence to the MedDiet was poor in 12.3%, average in 58.6%, and high in 29.1% of the subjects. Furthermore, 23.4% of patients were overweight/obese. The most impacting factors on BMI-SDS were skipping breakfast and their father's BMI. HbA1c and time in range % were positively associated with sweets and fish intake, respectively. Additionally, the father's socio-economic status (SES) and mother's age were associated with glucose control. Blood pressure was associated with travelling to school in vehicles, extra-virgin olive oil intake and milk/dairy consumption at breakfast. The promotion of the MedDiet, mainly having a healthy breakfast, is a good strategy to include in the management of T1D to improve glucose and metabolic control. This research is valuable for parents to obtain the best results for their children with T1D.

Keywords: diabetes; Mediterranean diet; pediatrics; food; nutrition; glucose; weight; blood pressure

Citation: Antoniotti, V.; Spadaccini, D.; Ricotti, R.; Carrera, D.; Savastio, S.; Goncalves Correia, F.P.; Caputo, M.; Pozzi, E.; Bellone, S.; Rabbone, I.; et al. Adherence to the Mediterranean Diet Is Associated with Better Metabolic Features in Youths with Type 1 Diabetes. *Nutrients* **2022**, *14*, 596. https://doi.org/10.3390/nu14030596

Academic Editor: Daniela Bonofiglio

Received: 7 January 2022
Accepted: 27 January 2022
Published: 29 January 2022

Publisher's Note: MDPI stays neutral with regard to jurisdictional claims in published maps and institutional affiliations.

1. Introduction

Type 1 diabetes (T1D) is an autoimmune disease that is rapidly increasing worldwide [1]. It is also one of the most common chronic diseases in childhood with an estimated prevalence of 1.1 million children and adolescents affected [2]. As well as insulin treatment, a crucial part of T1D patient therapy [3–5], another treatment is medical nutritional therapy (MNT). Its role is known in the management of type 2 diabetes, but it is also essential for T1D because of the need to adapt insulin therapy to nutrient intake, ensuring satisfactory glucose control [3,5,6]. Increased nutritional awareness of patients and their families is associated with better glucose control and the high quality of the diet in young people [7]. MNT assigned by a dietitian or a registered nutritionist is associated with a 1.0–1.9% decrease in Hba1c for people with T1D [8].

MNT includes recommendations for healthy eating patterns since patients with T1D are at high cardiovascular risk. The 2018 International Society for Pediatric and Adolescent

Diabetes (ISPAD) Guidelines recommend as proper nutrition habits for diabetes in children, a diet that allows for optimal growth, ideal weight maintenance and the prevention of acute and chronic complications of diabetes mellitus. The approximate energy intake and the essential nutrients should be distributed as follows: carbohydrates 45–55%, fats 30–35% and proteins 15–20% [5]. The bromatological composition is in line with that of the Mediterranean diet (MedDiet). In addition, there is evidence that low-carbohydrate diets can be nutritionally inadequate and may increase the risk of hypoglycemia [9]. Eating patterns like the MedDiet (based on wholegrain cereals, monounsaturated fats, plant-based food, reduced intake of red and processed meats) were suggested to be beneficial for long-term health, reducing cardiovascular risk [3,10]. The relationship between the MedDiet and diabetes mellitus has been widely studied, specifically related to type 2 diabetes mellitus (T2D) in adults, documenting beneficial effects in patients in terms of glucose control and primary and secondary prevention [11,12]. Several observational and some interventional studies demonstrated beneficial effects of the MedDiet on glucose homeostasis in individuals with T2D or metabolic syndrome. In a cross-sectional study involving 901 patients with T2D, the largest degree of adherence to the MedDiet was associated with low levels of hemoglobin A1C (HbA1c) and better post-prandial blood glucose control [13]. In addition to T2D, a multicenter observational study of 1076 pregnant women from ten Mediterranean countries, supported a reverse association between adherence to the MedDiet and the likelihood of gestational diabetes mellitus [14].

Data on nutritional habits in youths with T1D are scarce. Few studies have evaluated the effectiveness of the MedDiet in improving glucose control and the lipid profile. The main studies regarding T1D involved adult subjects. Observational data of the collaboration of European Childhood Diabetes Registers (EURODIAB) Prospective Complications Study examined the effects of saturated fatty acid (SFA) and dietary fibers on the onset of cardiovascular disease, showing an inverse correlation between fiber intake and cardiovascular risk in these patients [15]. Furthermore, the SEARCH Nutrition Ancillary Study confirmed that high adherence to the MedDiet was associated, overtime, with low levels of HbA1c and lipids in US youths [16]. These findings agree with an intervention study on a structured education about the MedDiet conducted in Italy by our group on adolescents with T1D [10]. The adherence to the MedDiet seemed more favorable than that to the Dietary Approaches to Stop Hypertension (DASH) diet since its impact was demonstrated only in a cross-sectional study [16,17].

In the context of the limited reported data, the purpose of this study was to assess the nutritional habits and physical activity in a group of multi-ethnic youths with T1D in North of Italy, surveyed in relation to their adherence to the MedDiet and metabolic parameters.

2. Subjects and Methods

2.1. Population and Clinical Parameters

The subjects were 65 consecutive youths aged between 9 and 18 years old recruited from outpatients attending the Division of Pediatrics of our Hospital between 2018 and 2021. The study was approved by the Ethical Committee of Novara (protocol number 143/17) and conformed to the guidelines of the European Convention of Human Rights and Biomedicine for Research in Children. Patients were diagnosed with T1D according to the American Diabetes Association (ADA) criteria [3]. We included multi-ethnic subjects with a T1D diagnosis between 2002 and 2020 and a time from the onset of at least six months. Any HbA1c levels, insulin treatment (continuous insulin or multiple daily injections) and body mass index (BMI) were considered. Inability to understand the Italian language or difficult engagement with the cultural mediator was the only exclusion criteria. Weight, height and blood pressure (BP) were measured by the medical staff. The BMI and BMI z-score were calculated. Weight was measured to the nearest 100 g by using an electronic scale, and height by a Harpenden stadiometer to the nearest mm. BMI was calculated as the ratio between weight (kg) and squared height (m^2). Normal weight, overweight and obesity were classified according to the International Obesity Task Force (IOTF) growth charts [18]. Blood pressure was evaluated as suggested by the National High Blood Pressure Educational Program (NHBPEP) Working Group of the American Academy of Pediatrics (AAP) [19]. Socio-economic status (SES) was assessed on the educational level of both parents and the type of work, classified as low, medium and high SES.

2.2. Dietary and Physical Activity Assessment

Adherence to the MedDiet was evaluated using the validated Italian version of the Mediterranean Diet Quality Index (KIDMED) score (Supplementary Table S1), which is composed of 16 dichotomous (positive/negative) items on eating habits [20]. The test is divided into four questions with negative connotations (-1) and twelve questions evaluated with a positive score (+1). A total score was calculated, ranging from 4 to 12. The assessment of the test was interpreted according to the following classification: low adherence (total score ≤ 3), average adherence (total score between 4 and 7), high adherence (total score ≥ 8). Items and scores are reported in Supplementary Table S1. The Italian version of the International Physical Activity Questionnaire for Adolescents (IPAQ-A) was used to assess the level of physical activity. It consists of four domains of physical activity reflecting on the activities of the previous seven days: school-related physical activity (including activities during physical education lessons and breaks), transport, housework and leisure physical activity [21]. The level of physical activity was measured by the Metabolic Equivalent of Task (MET). Both questionnaires were completed with a driven interview conducted by researchers who also collected anthropometric and social data of parents/caregivers.

2.3. Biochemical Evaluation and Glucose Monitoring Parameters

We collected biochemical data and glucose monitoring parameters of the routine clinical care. After overnight fasting, high-density lipoprotein (HDL), low-density lipoprotein(LDL), triglycerides, total cholesterol, glucose and HbA1c levels were measured. Daily insulin demand was calculated as both total IU/day and per kg/day. In 31 patients, additional values from a continuous glucose monitoring device (CGM) were also available. Outcomes of CGM were extracted, including average blood glucose (mg/dL) and standard deviation (SD), time in range (TIR) of optimal target (70–180 mg/dL) and variability classified as low (<70 mg/dL) and very low (<54 mg/dL), high (>180 mg/dL) and very high (>250 mg/dL). We considered as glucose imbalance Hba1c more than 8.5% and TIR lower than 70%, two validated cut-offs that represented approximately one-third of each group [22].

2.4. Statistical Analysis

All analyses were performed using the Statistical Package for the Social Sciences, Version 26.0 (SPSS Inc., Chicago, IL, USA). Descriptive characteristics were presented for each gender and full sample. After verification of the normal distribution of continuous variables through the Shapiro–Wilks test and PP plot (probability–probability plot), descriptive characteristics were shown as the mean + standard deviation, while qualitative variables were presented as frequencies or medians. The chi-squared test, the t-test for independent variables (means) and Mann–Whitney U-test (medians) were performed to study differences between genders and p-value was assessed for all tests. Correlation analysis followed the assumptions of Pearson r (parametric data) or Spearman's ρ (non-parametric data) depending on the characteristics of the correlated variables. Statistically significant correlations were expressed as r/ρ and p-value < 0.05 or < 0.01. Regression analysis of quantitative continuous dependent variables was conducted for a restricted portion of studied variables by creating regression models through a stepwise linear regression method. Independent variables were chosen from statistically significant correlated variables and by excluding multicollinearity factors or contributing factors. The qualitative regression of ordinal data was performed through multinomial logistic analysis instead. Best fitting model/models were determined automatically from the software. The confidence interval (CI) was set at 95%.

3. Results

The clinical data of the enrolled patients are shown in Table 1. All the 65 recruited patients (age: 15.1 ± 2.3 years) completed the questionnaires. Most of the patients were normal weight (71.9%), with a prevalence of overweight and obesity of 15.6% and 7.8%, respectively. All patients were in insulin treatment by the time of the visit, with 69% using the basal-bolus treatment and the remaining 31% using subcutaneous infusion. Regarding the management of the disease, only 21.9% of subjects were able to maintain a TIR% over 70%, and HbA1c was >8.5% in 26.2% of subjects. Data were also analyzed according to gender (32 males, 33 females). Males were taller than females and had also lower weight z score and standardized body mass index (BMI SDS). Females were more overweight and obese than males nearly to significance. Metabolic characteristics were similar between genders.

Table 1. Descriptive characteristics of sample.

		Full Sample	Boys	Girls	*p*-Value
Age (years)		15.07 ± 2.27	14.96 ± 2.07	15.18 ± 2.48	0.704
Years since T1D diagnosis (y)		6.15 ± 4.29	6.29 ± 4.64	6.03 ± 3.99	0.818
Weight (kg)		56.2 ± 13.22	56.71 ± 14.01	55.71 ± 12.61	0.762
Weight z-score (sd)		**−0.07 ± 1.16**	**−0.37 ± 1.17**	**0.22 ± 1.08**	**0.038**
Height (m)		**1.63 ± 0.11**	**1.66 ± 0.11**	**1.60 ± 0.09**	**0.025**
Height z-score (ds)		−0.00 ± 0.95	−0.12 ± 1.04	0.12 ± 0.85	0.319
BMI (kg/m^2)		21.08 ± 3.85	20.37 ± 3.88	21.80 ± 3.74	0.140
BMI-SDS (sd)		**−0.13 ± 1.20**	**−0.44 ± 1.23**	**0.18 ± 1.10**	**0.036**
BMI-IOTF	−2 (extremely underweight)	1 (1.6%)	1 (3.1%)	0	0.060 [1]
	−1 (underweight)	2 (3.1%)	2 (6.3%)	0	
	0 (normal weight)	46 (71.9%)	24 (75%)	22 (68.8%)	
	1 (overweight)	10 (15.6%)	3 (9.4%)	7 (21.9%)	
	2 (obese)	5 (7.8%)	2 (6.3%)	3 (9.4%)	
SBP (mmHg)		110.97 ± 10.10	110.48 ± 9.37	111.45 ± 10.91	0.226
DBP (mmHG)		69.34 ± 9.01	67.97 ± 8.69	70.71 ± 9.25	0.192
Mean Blood Glucose (mg/dL)		178.03 ± 32.25	180.80 ± 37.98	175.44 ± 26.80	0.651
HbA1c (%)		8.04 ± 1.78	8.16 ± 1.79	7.93 ± 1.79	0.609
TIR (%)		55.16 ± 18.54	53.42 ± 21.15	56.89 ± 16.02	0.604
UI insulin/kg/die		0.64 ± 0.26	0.64 ± 0.30	0.64 ± 0.23	0.894
Insulin treatment (1/2) [2]		1 = 45 (69.2%) 2 = 20 (30.8%)	1= 23 (71.9%) 2 = 9 (28.1%)	1 = 22(66.6%) 2 = 11(33.3%)	0.644
Total Cholesterol (mg/dL)		162.21 ± 28.89	159.13 ± 32.02	165.19 ± 25.69	0.417
HDL Cholesterol (mg/dL)		59.77 ± 15.23	59.73 ± 15.99	59.81 ± 14.74	0.984
Triglycerides (mg/dL)		71.03 ± 42.72	74.67 ± 50.54	67.62 ± 34.31	0.521
KIDMED (pts)		6 (−1/+12)	6 (2/12)	6 (−1/+11)	0.082 [1]

[1]: calculated with Mann–Whitney U-test. [2]: 1—basal/bolus injection/2—intravenous infuser. Descriptive characteristics are expressed as mean ± ds or median (min/max) or as frequencies (percentages). *p*-value expresses difference between genders. Statistically significant differences in the variables between genders are showed in bold. T1D—Type 1 diabetes, BMI—Body Mass Index, BMI-SDS—Standardized Body Mass Index, BMI-IOTF—International Obesity Taskforce, SBP—systolic blood pressure, DPB—diastolic blood pressure, HbA1c—glycated hemoglobin, TIR—time in range, HDL—high density lipoprotein, KIDMED—Mediterranean Diet Quality Index.

The KIDMED average (median) score was lower than recommended in both genders, with an almost statistically significant difference between boys and girls (Δ = 1.15 pts,

$p = 0.082$) (Table 1). KIDMED scores showed that (≥8) 29.1% of the subjects adhered to an optimal Mediterranean diet, (4–7) 58.5%of the subjects needed to improve, and (≤3) 12.3% of the whole sample exhibited a low-quality diet (Table 2). Concerning KIDMED items, frequencies and statistics for both genders are shown in Table 2. More than 80% of the subjects were eating at least one portion of vegetables each day, and more than 60% were eating at least two portions. Adequate fish and legume consumption was found in approximately 50% of the subjects. Males were more likely to eat at least one fruit/day ($p < 0.01$), ate less processed food at breakfast ($p < 0.05$) and more dairy foods daily ($p < 0.05$) compared to females.

Table 2. KIDMED items and other nutrition-related statistics in the sample.

		Full Sample	Boys	Girls	*p*-Value
KIDMED class	1	9 (13.8%)	2 (6.3%)	7 (21.2%)	0.55
	2	37 (56.9%)	18 (56.3%)	19 (57.6%)	
	3	19 (29.2%)	12 (37.5%)	7 (21.2%)	
Celiac disease (Y/N)		Y = 14 (21.5%); N = 51 (78.5%)	Y = 7 (21.9%); N = 25 (78.1%)	Y = 7 (21.2%); N = 26 (78.8%)	0.949
CHO calculation (Y/N)		Y = 44 (67.7%); N = 21 (32.2%)	Y = 22 (68.8%); N = 10 (31.3%)	Y = 22 (66.7%); N = 11 (33.3%)	0.859
1 p. Fruit/day (0/+1)		**0 = 14 (21.5%); 1 = 51 (78.5%)**	**0 = 2 (6.3%); 1 = 30 (93.8%)**	**0 = 12 (36.4%); 1 = 21 (63.6%)**	**0.003**
2 p. Fruit/day (0/+1)		0 = 36 (55.4%); 1 = 29 (44.6%)	0 = 15 (46.9%); 1 = 17 (53.1%)	0 = 21 (63.6%); 1 = 12 (36.4%)	0.177
1 p. Vegetables/day (0/+1)		0 = 10 (15.4%); 1 = 55 (84.6%)	0 = 6 (18.8%); 1 = 26 (81.3%)	0 = 4 (12.1%); 1 = 29 (87.9%)	0.462
2 p. Vegetables/day (0/+1)		0 = 24 (36.9%); 1 = 41 (63.1%)	0 = 12 (37.5%); 1 = 20 (62.5%)	0 = 12 (36.4%); 1 = 21 (63.6%)	0.925
Fish/2 or 3 p. each week (0/+1)		0 = 29 (44.6%); 1 = 36 (55.4%)	0 = 14 (43.8%); 1 = 18 (56.3%)	0 = 15 (45.5%); 1 = 18 (54.5%)	0.891
Fast Food once a week (−1; 0)		−1 = 1 (1.5%); 0 = 64 (98.5%)	−1 = 1 (3.1%); 0 = 31 (96.9%)	0 = 33 (100%)	0.310
Legumes at least 1 p. a week (0; +1)		0 = 35 (53.8%); 1 = 30 (46.2%)	0 = 19 (59.4%); 1 = 13 (40.6%)	0 = 16 (48.5%); 1 = 17 (51.5%)	0.382
Cereals (pasta, rice . . .) at least 5 p. a week (0; +1)		0 = 2 (3.1%); 1 = 63 (96.9%)	1 = 32 (100%)	0 = 2 (6.1%); 1 = 31 (93.9%)	0.160
Cereals at breakfast (0; +1)		0 = 46 (70.8%); 1 = 19 (29.2%)	0 = 20 (62.5%); 1 = 12 (37.5%)	0 = 26 (78.8%); 1 = 7 (21.2%)	0.152
Nuts and similar foods 2 or 3 p. a week (0; +1)		0 = 48 (73.8%); 1 = 17 (26.2%)	0 = 21 (65.6%); 1 = 11 (34.4%)	0 = 27 (81.8%); 1 = 6 (18.2%)	0.141
Olive Oil as preferred oil (0; +1)		0 = 1 (1.5%); 1 = 64 (98.5%)	1 = 32 (100%)	0 = 1 (3%); 1 = 32 (97%)	0.325
Skip breakfast (−1; 0)		−1 = 7 (10.8%); 0 = 58 (89.2%)	−1 = 3 (9.4%); 0 = 29 (90.6%)	−1 = 4 (12.1%); 0 = 29 (87.9%)	0.723
Milk/yogurt or dairy food at breakfast (0; +1)		0 = 18 (27.7%); 1 = 47 (72.3%)	0 = 8 (25%); 1 = 24 (75%)	0 = 10 (30.3%); 1 = 23 (69.7%)	0.636
Processed Food at breakfast (−1; 0)		**−1 = 53 (81.5%); 0 = 12 (18.5%)**	**−1 = 22 (68.8%); 0 = 10 (31.3%)**	**−1 = 31 (93.9%); 0 = 2 (6.1%)**	**0.034**
2 portions of milk/yogurt or dairy foods/day (0; +1)		**0 = 29 (44.6%); 1 = 36 (55.4%)**	**0 = 10 (31.3%); 1 = 22 (68.8%)**	**0 = 19 (57.6%); 1 = 14 (42.4%)**	**0.034**
Sweets and/or candies every day (−1; 0)		−1 = 20 (30.8%); 0 = 45 (69.2%)	−1 = 11 (34.4%); 0 = 21 (65.6%)	−1 = 9 (27.3%); 0 = 24 (72.7%)	0.538

KIDMED class: 1—low-quality diet; 2 KIDMED class: 1—to improve; 3—optimal Mediterranean diet. (0/+1): 0—No, +1—Yes. (−1; 0): −1—Yes, 0—No. Statistically significant differences in the variables between genders are showed in bold.

Other characteristics of the sample, including the socio-economic aspects of families and the main physical activity items assessed through IPAQ-A, are presented in Table 3. The IPAQ-A additional items' descriptive characteristics are shown in Supplementary Table S2. According to our data, mild physical activity was slightly more appreciated by the females, while vigorous physical activity was higher in males, even if the difference was not statistically significant (p = ns). Family count varied from two to six individuals with four individuals as the most frequent case (45.9%). The SES of the mother and father were not different, neither was the smoking habits of parents even if the fathers had twice the frequency of smoking (29.5%) compared to the mothers (14.8%).

Table 3. IPAQ-A global assessments and other non-nutritional aspects of sample.

		Full Sample	Boys	Girls	*p*-Value
Family count	2	2 (3.3%)	1 (3.6%)	1 (3%)	0.138
	3	13 (21.3%)	9 (32.1%)	4 (12.1%)	
	4	28 (45.9%)	11 (39.3%)	17 (51.5%)	
	5	12 (19.7%)	5 (17.9%)	7 (21.2%)	
	6	6 (9.8%)	2 (7.1%)	4 (12.1%)	
SES mother *	1	27 (42.2%)	13 (40.6%)	14 (43.8%)	0.623
	2	31 (48.4%)	15 (46.9%)	16 (50%)	
	3	6 (9.4%)	4 (12.5%)	2 (6.3%)	
SES father *	1	36 (56.3%)	19 (59.4%)	17 (53.1%)	0.564
	2	25 (39.1%)	12 (37.5%)	13 (40.6%)	
	3	3 (4.7%)	1 (3.1%)	2 (6.3%)	
Mother smoke (Y/N) **		Y = 9 (14.8%); N = 52 (85.2)%	Y = 5 (17.2%); N = 24 (82.9%)	Y = 4 (12.1%); N = 28 (84.8%)	0.602
Father smoke (Y/N) **		Y = 18 (29.5%); N = 43 (70.5%)	Y = 8 (27.6%); N = 21 (72.4%)	Y = 10 (31.3%); N = 22 (68.8%)	0.756
3.3 MET (kcal/week)		809.01 ± 844.37	687.33 ± 626.77	927.00 ± 1008.07	0.331
4 MET (kcal/week)		590.77 ± 760.65	654.38 ± 908.70	529.09 ± 591.00	0.535
8 MET (kcal/week)		817.23 ± 1045.97	975.00 ± 1105.19	664.24 ± 977.65	0.259
Total kcal burnt/week		2217.01 ± 1341.71	2316.70 ± 1272.58	2120.33 ± 1418.43	0.348

SES—Socio-economic status; 3.3 MET—kcal burnt with walking/mild physical activity in the last week; 4 MET—kcal burnt with moderate physical activity in the last week; 8 MET—kcal burnt with vigorous physical activity in the last week; *—difference between mother and father assessed through X^2 test: $p = 0.181$; **—difference between mother and father assessed through X^2 test: $p = 0.063$.

3.1. Correlations and Regressions

3.1.1. Weight and BMI

The evaluation of significant correlations between organic/anthropometric parameters and non-biological/nutritional/physical activity habits has provided a lot of data which are summarized in Supplementary Tables S3–S5. The weight z-score was correlated with a higher family count ($\rho = 0.321$; $p < 0.05$), skipping breakfast ($\rho = -0.407$; $p = 0.001$) and with the weight and BMI of parents. BMI-SDS kept the same tendency but added a direct correlation with legume consumption ($\rho = 0.262 < 0.05$). Increased IOTF-BMI was also correlated with a reduced intake of nuts ($\rho = -0.425$; $p < 0.001$), cereals ($\rho = -0.248$, <0.05), dairy foods at breakfast ($\rho = -0.306$; $p < 0.05$) and lower 8 MET activities (-0.249; $p < 0.05$). Stepwise regression (Table 4) showed that, for the whole sample, the most impacting independent factor on BMI-SDS was skipping breakfast, followed by the father's BMI. Considering the differences between genders, BMI-SDS was primarily associated with skipping breakfast and father's weight in females, while in males the main predicting factor was the time spent in physical activity during school even if skipping breakfast, reduced nut consumption and a higher family count ($p < 0.05$) were found to be correlated with increased BMI-SDS in males. IOTF-BMI class was primarily associated with nut consumption and 8 MET above all other effects.

Table 4. Multiple stepwise regression analysis results of significant effects (independent variables) on BMI-SDS (dependent variable) and multiple logistic stepwise regression on IOTF-BMI.

Dependent Variables	Significant Effects	B (95% CI)	β	*p* Value
BMI-SDS (sd): Model 1	Skip breakfast (−1/0)	−1.829 (−2.746; −0.912)	−0.471	<0.001
BMI-SDS (sd): Model 2	Skip breakfast (−1/0)	−1.804 (−2.675; −0.932)	−0.464	<0.001
	Father's BMI (kg/m^2)	0.074 (0.018; 0.129)	0.299	0.01
IOTF-BMI (pts) *	Nut consumption (0/1)	/	/	0.001
	8 MET	/	/	<0.05
Boys: BMI-SDS (sd)	Minutes of sport at school (min)	−0.016 (−0.027; −0.006)	−0.549	0.003
Females: BMI-SDS (sd): Model 1	Father's weight (kg)	0.033 (0.011; 0.054)	0.507	0.004
Females: BMI-SDS (sd): Model 2	Father's weight (kg)	0.027 (0.007; 0.047)	0.424	0.007
	Skip breakfast (−1/0)	−1.377 (−2.459; −0.295)	−0.396	0.015

CI—confidence interval. *—difference between mother and father assessed through X^2 test: $p = 0.181$.

3.1.2. Glucose Control

Mean blood glucose was correlated with a lower age of the mother ($r = -0.480$; $p < 0.01$), the SES of father ($\rho = -0.516$; $p < 0.01$), fish consumption ($\rho = -0.269$; $p < 0.05$) and cereal consumption at breakfast ($\rho = -0.418$; $p < 0.05$). No aspects of physical activity were found to influence mean glucose. Beyond the correlations found for fasting blood glucose, HbA1c was also correlated with a higher consumption of sweets ($\rho = -0.347$; $p < 0.01$) and less days and time spent on a bike ($\rho = -0.266$ and $\rho = -0.276$ respectively; $p < 0.05$).TIR%, which was highly and inversely correlated with HbA1c ($\rho = -0.881$; $p < 0.001$), as expected, kept almost the same significant correlations of the latter: higher fish consumption ($\rho = 0.390$; $p < 0.05$) and cereals at breakfast ($\rho = 0.427$; $p < 0.05$), higher age of the mother ($r = 0.401$; $p < 0.05$), and the SES of the father ($\rho = 0.471$; $p < 0.01$).Stepwise regression (Table 5) conducted on these three main indicators of glucose control revealed that mean blood glucose was primarily predicted by the age of the mother and the SES of the father, while HbA1c was mainly associated with the SES of the father and the consumption of sweets. TIR% was instead even more associated with fish consumption, and its relationship with the SES of the father was revealed to be multiplicatively and independently associated with TIR% as the first model of regression. The KIDMED score was found only inversely correlated with the risk of hyperglycemia ($\rho = 0.387$; $p < 0.05$).

Table 5. Multiple stepwise regression analysis results of significant effects (independent variables) on glucose control main parameters (dependent variables).

Dependent Variables	Significant Effects	B (95% CI)	β	*p* Value
Mean blood glucose (mg/dL) Model 1	Mother's age (y)	−3.321 (−5.628; −1.015)	−0.480	0.006
Mean blood glucose (mg/dL) Model 2	Mother's age (y)	−2.851 (−4.976; −0.727)	−0.412	0.010
	Father's SES (1/2/3)	−20.834 (−36.655; −5.013)	−0.404	0.012

Table 5. *Cont.*

Dependent Variables	Significant Effects	B (95% CI)	β	*p* Value
HbA1c (%) Model 1	Father's SES (1/2/3)	−1.050 (−1.808; −0.291)	−0.345	0.008
HbA1c (%) Model 2	Father's SES (1/2/3)	−1.106 (−1.817; −0.395)	−0.363	0.003
	Consumption of Sweets (−1/0)	−1.379 (−2.295; −0.464)	−0.352	0.004
TIR (%) Model 1	Father's SES (1/2/3)	15.718 (6.230; 25.206)	0.564	0.002
TIR (%) Model 2	Father's SES (1/2/3)	14.316 (5.403; 23.228)	0.513	0.003
	Fish consumption (0/1)	12.327 (1.014; 23.640)	0.348	<0.05

3.1.3. Blood Pressure and Others

Blood pressure was found to be mainly linked with breakfast habits. Diastolic blood pressure (DBP) was correlated with less cereal consumption in the morning ($\rho = -0.264$; $p < 0.05$) and the consequently higher processed products consumed at breakfast ($\rho = 0.260$; $p < 0.05$). In addition, DPB was inversely correlated with extra-virgin olive oil intake ($\rho = -0.296$; $p < 0.05$). Systolic blood pressure (SPB) correlated with less milk/dairy foods at breakfast instead ($\rho = -0.291$; $p < 0.05$) and almost significantly with extra virgin olive oil consumption ($p < 0.057$). Physical activity items were found to be partially correlated with blood pressure, in particular traveling by car or public transport was correlated with both DBP ($\rho = 0.338$; $p < 0.01$) and SBP ($\rho = 0.301$; $p < 0.01$). Subsequent stepwise regression analysis (Table 6) confirmed that DBP was firstly associated with motorized travels for reaching school, followed by the other factors, while SPB was mainly predicted by olive oil and milk/dairy consumption at breakfast. Correlation analysis also showed that blood lipids were not linked to any of the nutritional and physical activity items (p = ns).

Table 6. Multiple stepwise regression analysis results of significant effects (independent variables) on blood pressure (dependent variables).

Dependent Variables	Significant Effects	B (95% CI)	β	*p* Value
SBP (mmHg) Model 1	Milk/Dairy food at breakfast (0/1)	−6.073 (−11.551; −0.595)	−0.275	0.030
SBP (mmHg) Model 2	Milk/Dairy food at breakfast (0/1)	−6.557 (−11.884; −1.230)	−0.297	0.017
	Olive oil (0/1)	−21.279 (−40.475; −2.083)	−0.268	0.030
DBP (mmHg) Model 1	Days to school with motorized transport (n)	1.764 (0.538; 2.991)	0.348	0.006
DBP (mmHg) Model 2	Days to school with motorized transport (n)	1.779 (0.601; 2.957)	0.351	0.004
	Cereals at breakfast (0/1)	−5.567 (−10.072; −1.061)	−0.287	0.016
DBP (mmHg) Model 3	Days to school with motorized transport (n)	1.560 (0.396; 2.724)	0.308	0.009
	Cereals at breakfast (0/1)	−7.005 (−11.592; 2.418)	−0.361	0.003
	Processed Food at breakfast (0/1)	5.738 (0.303; 11.173)	0.254	0.039
DBP (mmHg) Model 4	Days to school with motorized transport (n)	1.342 (0.194; 2.490)	0.265	0.023
	Cereals at breakfast (0/1)	−6.703 (−11.166; −2.241)	−0.346	0.004
	Processed Food at breakfast (0/1)	6.144 (0.854; 11.434)	0.272	0.024
	Olive oil (0/1)	−16.997 (−32.881; −1.114)	−0.240	0.036

4. Discussion

MNT is a crucial element to improve glucose control and reduce cardiovascular risk in T1D in youths. Among dietary patterns, increasing evidence supports the MedDiet, mainly in adults with T2D [23], due its antioxidant and anti-inflammatory properties; however, results in the pediatric age should be implemented. This is one of the few studies that has aimed to describe the relationship among the components of the MedDiet and global metabolic control in children and adolescents with T1D.

4.1. Dietary Factors Associated with the Risk of Obesity

We observed that the prevalence of youths with T1D being overweight and obese is similar to the healthy pediatric population living in Italy [24–26] who also follow as an average adherence to the MedDiet [27,28]. Similar findings on the weight status of children with T1D have been reported in other ethnic groups or countries corroborating the general global trend [29–34]. Furthermore, females in our cohort were more at risk similar to other reports [32,33,35]. In all these studies, several factors have been associated with this risk of overweight/obesity in T1D as increased functional growth hormone (GH) secretion, longer disease duration, intensive insulin therapy in relation to pubertal insulin-resistance or flexible eating patterns, high insulin doses that could inhibit protein catabolism and slow basal metabolism or frequent snacking to avoid hypoglycemia [36,37]. However, only a few authors focused on dietary risk factors and even fewer on the MedDiet. Two US studies conducted using 287 pediatric subjects aged between 8 and 21 years old showed that the obesity risk was related to a frequent intake of unhealthy foods poor in fibers and micronutrients [38,39]. Interestingly, we observed that increased weight is associated with breakfast skipping habits, confirmed by the general pediatric population, as recently reviewed by our group in both cross-sectional and intervention trials [40,41]. The effect of skipping the first morning meal, despite being made up by similar calories in the day, is complex and partially unexplained, although several mechanisms have been hypothesized, such as circadian misalignment, the length of the night fasting and other unhealthy food habits. However, the confirmation in children and adolescents with T1D suggests that the alteration of the chronotype is a key feature of obesity development in any condition. Furthermore, we observed that the intake of dairy products at breakfast was also negatively associated with the risk of being overweight and obese. With the urbanization of people living in the Mediterranean area, youth are deviating to a Western diet rich in saturated fat, simple carbohydrates and refined and processed foods. One of the typical phenomena of this nutrition transition is the choice of sugar-sweetened beverages at breakfast instead of dairy products [42]. Other studies have recently reported this association in children with obesity [43–47]. Behind the excess of simple sugars instead of foods rich in nutrients, several other mechanisms have been reported, such as derangements in the calcium homeostasis, insulin secretion, satiety and satiation and gut-microbiota when the intake of dairy products is low [45,48,49]. In our cohort a low intake of nuts and cereals was associated with an increased risk of being overweight and obese. Two studies on adults with T1D, the Finnish Diabetic Nephropathy (FinnDiane) Study of 1058 individuals [50] and the EURODIAB Complications Study of 2868 individuals [51], confirmed our pediatric results on nuts and cereals, respectively. These findings could hide a high consumption of unhealthy processed foods rich in saturated and trans fats, and sugars instead of nuts and cereals. Although nuts are dense-energy foods, they are rich in micronutrients and bio-active compounds. Growing evidence suggests their role in protecting from obesity or helping in weight loss [52–54] due to several mechanisms including prolonged chewing, delayed gastric emptying because of fibers and unsaturated fats, increased satiety and booster actions on lipid oxidation and thermogenesis [55].

4.2. Dietary Factors Associated with the Glucose Control

Although a better KIDMED score only reduced the risk of hyperglycemia differently from a modified version that correlated with HbA1c [16], low mean glucose levels, HbA1c and high TIR% were all associated with determinants of the KIDMED score, such as a high consumption of fish, cereals at breakfast, and a low consumption of sweets and candies. As previously discussed for weight, foods rich in nutrients and maintenance of breakfast habits are confirmed as being correlated with better glucose control, diversely from foods rich in saturated fats and sugars. Another study previously discussed on weight, observed that poor diabetic control was associated with unhealthy foods [39], as well as another one in Brazilian children with ultra-processed foods, a category that also comprises sweets and candies [56]. Although fasting glucose levels, HbA1c and TIR% were all correlated and associated with the same factors, interestingly, the main dietary determinants in the regression models were different. A high HbA1c was mostly related to a high intake of sweets which are rich in sugars. This is consistent with previous data on the fact that apart from total carbohydrate daily content, source of carbohydrate intake, and, therefore, glycemic load/index, is associated with HbA1c levels [57–60]. Furthermore, we observed that a high TIR was associated with a high intake of fish. No similar results have been published yet, whereas a better TIR has been recently associated with a carbohydrate intake range from 40–44% [61], or fish intake with a lower risk of microalbuminuria [62]. Fish is rich in macronutrients and micronutrients which are implicated with better metabolic control as well as low insulin resistance [63]. However, the human diet is complex, and different nutrients could have a synergistic effect. Because concluding data on the specific role of omega 3, vitamin D, unsaturated fats and proteins alone are still not conclusive concerning glucose control, and even less on TIR [64–66], further studies or post-hoc analysis from registries on T1D are needed.

4.3. Dietary Factors Associated with Blood Pressure

In line with findings on glucose controls, blood pressure was associated with breakfast habits. The high consumption of cereals and milk/dairy foods, and low consumption of processed products at breakfast were associated with better DBP and SBP levels.All these components of breakfast mirror healthy habits balanced in nutrients and are typical not only of the MedDiet but also of the DASH diet [67,68]. Adherence to it has been demonstrated to reduce the risk of hypertension in the SEARCH for Diabetes in Youth Study [69]. The effect on blood pressure dueto the consumption of processed foods is likely due to several factors, but surely one of the most important factors is the contribution of increasedsaltintake [70]. Regarding milk products, the relatively high potassium content of milk and dairy products implies that increased intake of them may reduce blood pressure, as also observed with studies on the DASH diet or lacto-ovo vegetarian diet [49,67]. This seems more pronounced when dairy products are consumed instead of juice or sugar-sweetened beverages [49,71,72]. No reports are present for T1D, distinct from T2D or metabolic syndrome [67,73], and our data strengthen the evidence in other diseases.

Extra virgin olive oil (EVOO) reduces cardiovascular risk in adults as demonstrated by numerous studies, including the Nurses' Health Study II, the Health Professional's Follow-up Study and the Prevention with Mediterranean Diet (PREDIMED) Study [74]. Furthermore, both experimental and human observational and intervention studies demonstrated EVOO has anti-hypertensive actions due to its chemical composition, mainly characterized by oleic acid, polyphenols and other antioxidant compounds [75]. Our data on EVOO and DBP are in line with a recent intervention trial in adults with T1D in which EVOO improved vascular function [76].

4.4. Family, Social Factors and Physical Activity

Interestingly, we observed that BMI-SDS and glucose control were associated directly with the BMI of the father, the latter indirectly with the SES of the father and the mother's age. It is well known that weight in offspring is partly inherited by parents but also due to the family setting, in particular the SES. All of these are markers of a complex interplay among genetics and environment that have the most significant impact on children younger than ten years old. However, data on the role of paternal BMI and mothers' age is scarce due to the lack of studies for these factors [77–79]. Indeed, our findings strengthen some data in these subjects with regards to T1D. Regarding the role of the family on offspring concerning glucose control, while SES is associated with clear evidence, parental characteristics need further study because the role they play is inconclusive [80]. However, regarding the mother's crucial involvement in T1D management [80], we can speculate that maternal age could result in different maternal parenting styles and coping abilities [81,82]. We observed some associations among physical activity determinants and weight and diastolic blood pressure. Most of the results are related to sedentary behaviors, such as going to school by car, which likely reflects a generally reduced physical activity that could be underreported. Our results strengthen the importance of empowerment to improve physical activity in youths with T1D to reduce cardiovascular risk [5,83].

4.5. Study Limitations and Strengths

This study has some limitations. First, it was a cross-sectional study design, and we could not establish causal relationships between MedDiet adherence and health outcomes. Second, we used the KIDMED score without integrating it with a food frequency questionnaire. However, periodically, our patients carried out a dietary survey by dieticians for the improvement of carbohydrate counting, indeed, they were educated about the assessment of diet quality. Furthermore, the KIDMED score is the most used index of adherence in the pediatric literature and by using it in the past we have obtained interesting results, and it has been validated through use on the general population [27,28]. Indeed, we cannot exclude further nutrients from being players behind the food habits we reported in T1D.Unexpectedly, we failed to show any correlation with fruit or vegetable intake. This could result from high consumption of fruits and vegetables in our cohort, likely due to the dietary counseling for T1D. Furthermore, physical activity had minimal impact on the features we analyzed. Although the International Physical Activity Questionnaire (IPAQ) was validated on children and adolescents [21], movement activities linked to free play among peers could be underreported. In general, we demonstrated that good adherence to the MedDiet, in particular to some food habits, was associated with a better metabolic status in youth with T1D.

5. Conclusions

Dietary risk factors, typical of low adherence to the MedDiet, were associated with a high risk of obesity in T1D, which were the same as the general pediatric and adult population. This suggests that youths with T1D are not protected from increased weight whether they continuously adhere to healthy food patterns or not. The habit of having breakfast with foods rich in nutrients, including dairy products, is one of the most important dietary features we found associated with good glucose and metabolic control. The promotion of breakfast, as well as discouraging the excessive intake of ultra-processed foods, are confirmed key determinants in the prevention of negative health effects of nutrition transition in youth with T1D. Even if the cardiovascular protective role of EVOO is reaching amassing evidence in children, more attention and further research are needed to understand the role of fish consumption in glucose regulation. The promotion of the MedDiet is a good strategy to include in the management of youth with T1D, and all parents should be aware of this to obtain the best results for their children in the management of the disease.

Supplementary Materials: The following are available online at https://www.mdpi.com/article/10.3390/nu14030596/s1, Table S1: KIDMED items, Table S2: Descriptive characteristics of specific IPAQ-A (Italian version) items in the sample, Table S3: Correlations between clinical characteristics and KIDMED items, Table S4: Correlations between clinical characteristics and IPAQ-A items, Table S5: Correlations between clinical characteristics of children and mother and father data.

Author Contributions: Conceptualization, F.P. and S.B.; Methodology, V.A., D.S., R.R., D.C., F.P. and F.P.G.C.; Software, D.S.; Validation, R.R. and M.C.; Formal Analysis, V.A., D.S. and F.P.; Investigation, V.A., R.R., E.P., D.C. and S.S.; Data Curation, M.C.; Writing Original Draft Preparation, V.A., D.S. and F.P.; Writing Review and Editing, S.B. and I.R.; Supervision, S.B., I.R. and F.P.; Funding Acquisition, F.P. All authors discussed the results and contributed to the final manuscript. All authors affirm that the present work is original, has not been published previously and has not been submitted elsewhere for consideration of print or electronic publication. Each person listed as an author participated in the work in a substantive manner, in accordance with ICMJE authorship guidelines, and is prepared to take public responsibility for it. All authors have read and agreed to the published version of the manuscript.

Funding: This research was partially supported by: Ministry of Education, Universities, and Research: Grant PRIN (2020NCKXBR_004); Ministry of Education, Universities, and Research, Department of Excellence grant (FOHN project, KETOMI project).

Institutional Review Board Statement: The study was conducted according to the guidelines of the Declaration of Helsinki, approved by the Ethical Committee of Novara (protocol number 143/17) and conformed to the guidelines of the European Convention of Human Rights and Biomedicine for Research in Children.

Informed Consent Statement: Informed consent was obtained from all subjects involved in the study.

Data Availability Statement: The data presented in this study are contained within the article and the Supplementary Materials.

Acknowledgments: The authors thank all the children and their parents.

Conflicts of Interest: The authors declare no conflict of interest.

References

1. Dabelea, D. The accelerating epidemic of childhood diabetes. *Lancet* **2009**, *373*, 1999–2000. [CrossRef]
2. IDF Diabetes Atlas. Available online: https://www.idf.org/e-library/epidemiology-research/diabetes-atlas/159-idf-diabetes-atlas-ninth-edition-2019.html (accessed on 17 December 2021).
3. American Diabetes Association. Standards of medical care in diabetes, 2021. *Diabetes Care* **2020**, *44* (Suppl. 1), S1–S2.
4. Società Italiana di Diabetologia (SID). Standard Italiani per la Cura del Diabete Mellito. Available online: https://www.siditalia.it/pdf/StandarddiCuraAMD-SID2018_protetto2.pdf (accessed on 14 December 2021).
5. Smart, C.E.; Annan, F.; Higgins, L.A.; Jelleryd, E.; Lopez, M.; Acerini, C.L. ISPAD Clinical Practice Consensus Guidelines 2018: Nutritional management in children and adolescents with diabetes. *Pediatr. Diabetes* **2018**, *19*, 136–154. [CrossRef] [PubMed]
6. Mehta, S.N.; Volkening, L.K.; Anderson, B.J.; Nansel, T.; Weissberg-Benchell, J.; Wysocki, T.; Laffel, L.M.B.; Steering Committee for the Family Management of Childhood Diabetes Study. Dietary Behaviors Predict Glycemic Control in Youth With Type 1 Diabetes. *Diabetes Care* **2008**, *31*, 1318–1320. [CrossRef] [PubMed]
7. Rovner, A.J.; Nansel, T.R.; Mehta, S.N.; Higgins, L.A.; Haynie, L.D.; Laffel, L.M. Development and Validation of the Type 1 Diabetes Nutrition Knowledge Survey. *Diabetes Care* **2012**, *35*, 1643–1647. [CrossRef] [PubMed]
8. Franz, M.J.; MacLeod, J.; Evert, A.; Brown, C.; Gradwell, E.; Handu, D.; Reppert, A.; Robinson, M. Academy of Nutrition and Dietetics Nutrition Practice Guideline for Type 1 and Type 2 Diabetes in Adults: Systematic Review of Evidence for Medical Nutrition Therapy Effectiveness and Recommendations for Integration into the Nutrition Care Process. *J. Acad. Nutr. Diet.* **2017**, *117*, 1659–1679. [CrossRef]
9. Ranjan, A.; Schmidt, S.; Damm-Frydenberg, C.; Steineck, I.; Clausen, T.R.; Holst, J.J.; Madsbad, S.; Nørgaard, K. Low-Carbohydrate Diet Impairs the Effect of Glucagon in the Treatment of Insulin-Induced Mild Hypoglycemia: A Randomized Crossover Study. *Diabetes Care* **2017**, *40*, 132–135. [CrossRef]
10. Cadario, F.; Prodam, F.; Pasqualicchio, S.; Bellone, S.; Bonsignori, I.; Demarchi, I.; Monzani, A.; Bona, G. Lipid profile and nutritional intake in children and adolescents with Type 1 diabetes improve after a structured dietician training to a Mediterranean-style diet. *J. Endocrinol. Investig.* **2012**, *35*, 160–168. [CrossRef]
11. Toobert, D.J.; Glasgow, R.E.; Strycker, L.A.; Barrera, M.; Radcliffe, J.L.; Wander, R.C.; Bagdade, J.D. Biologic and quality-of-life outcomes from the Mediterranean Lifestyle Program: A randomized clinical trial. *Diabetes Care* **2003**, *26*, 2288–2293. [CrossRef]
12. Huo, R.; Du, T.; Xu, Y.; Xu, W.; Chen, X.; Sun, K.; Yu, X. Effects of Mediterranean-style diet on glycemic control, weight loss and cardiovascular risk factors among type 2 diabetes individuals: A meta-analysis. *Eur. J. Clin. Nutr.* **2015**, *69*, 1200–1208. [CrossRef]
13. Esposito, K.; Maiorino, M.I.; Di Palo, C.; Giugliano, D.; Group for the Campanian Postprandial Hyperglycemia Study. Adherence to a Mediterranean diet and glycaemic control in Type 2 diabetes mellitus. *Diabet. Med.* **2009**, *26*, 900–907. [CrossRef] [PubMed]

14. Karamanos, B.; Thanopoulou, A.; Anastasiou, E.; Assaad-Khalil, S.; Albache, N.; Bachaoui, M.; Slama, C.B.; El Ghomari, H.; Jotic, A.; Lalic, N.; et al. Relation of the Mediterranean diet with the incidence of gestational diabetes. *Eur. J. Clin. Nutr.* **2014**, *68*, 8–13. [CrossRef] [PubMed]
15. Schoenaker, D.A.J.M.; Toeller, M.; Chaturvedi, N.; Fuller, J.H.; Soedamah-Muthu, S.S.; EURODIAB Prospective Complications Study Group. Dietary saturated fat and fibre and risk of cardiovascular disease and all-cause mortality among type 1 diabetic patients: The EURODIAB Prospective Complications Study. *Diabetologia* **2012**, *55*, 2132–2141. [CrossRef] [PubMed]
16. Zhong, V.W.; Lamichhane, A.P.; Crandell, J.L.; Couch, S.C.; Liese, A.D.; The, N.S.; Tzeel, B.A.; Dabelea, D.; Lawrence, J.M.; Marcovina, S.M.; et al. Association of adherence to a Mediterranean diet with glycemic control and cardiovascular risk factors in youth with type I diabetes: The SEARCH Nutrition Ancillary Study. *Eur. J. Clin. Nutr.* **2016**, *70*, 802–807. [CrossRef] [PubMed]
17. Barnes, T.L.; Crandell, J.L.; Bell, R.A.; Mayer-Davis, E.J.; Dabelea, D.; Liese, A.D. Study Group for the SEARCH for Diabetes in Youth. Change in DASH diet score and cardiovascular risk factors in youth with type 1 and type 2 diabetes mellitus: The SEARCH for Diabetes in Youth Study. *Nutr. Diabetes* **2013**, *3*, e91. [CrossRef] [PubMed]
18. Cole, T.J.; Lobstein, T. Extended international (IOTF) body mass index cut-offs for thinness, overweight and obesity. *Pediatr. Obes.* **2012**, *7*, 284–294. [CrossRef] [PubMed]
19. Flynn, J.T.; Kaelber, D.C.; Baker-Smith, C.M.; Blowey, D.; Carroll, A.E.; Daniels, S.R.; de Ferranti, S.D.; Dionne, J.M.; Falkner, B.; Flinn, S.K.; et al. Clinical Practice Guideline for Screening and Management of High Blood Pressure in Children and Adolescents. *Pediatrics* **2017**, *140*, e20171904. [CrossRef] [PubMed]
20. Serra-Majem, L.; Ribas, L.; Ngo, J.; Ortega, R.M.; García, A.; Pérez-Rodrigo, C.; Aranceta, J. Food, youth and the Mediterranean diet in Spain. Development of KIDMED, Mediterranean Diet Quality Index in children and adolescents. *Public Health Nutr.* **2004**, *7*, 931–935. [CrossRef]
21. Hagströmer, M.; Bergman, P.; De Bourdeaudhuij, I.; Ortega, F.B.; Ruiz, J.R.; Manios, Y.; Rey-López, J.P.; Phillipp, K.; von Berlepsch, J.; Sjöström, M.; et al. Concurrent validity of a modified version of the International Physical Activity Questionnaire (IPAQ-A) in European adolescents: The HELENA Study. *Int. J. Obes.* **2008**, *32*, S42–S48. [CrossRef]
22. Battelino, T.; Danne, T.; Bergenstal, R.M.; Amiel, S.A.; Beck, R.; Biester, T.; Bosi, E.; Buckingham, B.A.; Cefalu, W.T.; Close, K.L.; et al. Clinical Targets for Continuous Glucose Monitoring Data Interpretation: Recommendations From the International Consensus on Time in Range. *Diabetes Care* **2019**, *42*, 1593–1603. [CrossRef]
23. Esposito, K.; Maiorino, M.I.; Bellastella, G.; Chiodini, P.; Panagiotakos, D.; Giugliano, D. A journey into a Mediterranean diet and type 2 diabetes: A systematic review with meta-analyses. *BMJ Open* **2015**, *5*, e008222. [CrossRef] [PubMed]
24. Nadrone, P.; Spinelli, A.; Buoncristiano, M.; Lauria, L.; Pierannunzio, D.; Galeone, D. Centro Nazionale per la Prevenzione delle Malattie e la Promozione della Salute. Okkio alla Salute: I Dati Nazionali 2016. Available online: https://www.epicentro.iss.it/okkioallasalute/Dati2016 (accessed on 30 July 2018).
25. Spinelli, A.; Nardone, P. Centro Nazionale per la Prevenzione delle Malattie e la Promozione della Salute. Okkio alla Salute: I Risultati dell'Indagine 2019 in Liguria. Available online: https://www.epicentro.iss.it/okkioallasalute/indagine-2019-report-liguria (accessed on 30 July 2018).
26. Roser, M.; Ritchie, H. Burden of Disease. 2016. Available online: https://ourworldindata.org/burden-of-disease (accessed on 12 December 2021).
27. Iaccarino Idelson, P.; Scalfi, L.; Valerio, G. Adherence to the Mediterranean Diet in children and adolescents: A systematic review. *Nutr. Metab. Cardiovasc. Dis.* **2017**, *27*, 283–299. [CrossRef] [PubMed]
28. Archero, F.; Ricotti, R.; Solito, A.; Carrera, D.; Civello, F.; Di Bella, R.; Bellone, S.; Prodam, F. Adherence to the Mediterranean Diet among School Children and Adolescents Living in Northern Italy and Unhealthy Food Behaviors Associated to Overweight. *Nutrients* **2018**, *10*, 1322. [CrossRef]
29. Oza, C.; Khadilkar, V.; Karguppikar, M.; Ladkat, D.; Gondhalekar, K.; Shah, N.; Khadilkar, A. Prevalence of metabolic syndrome and predictors of metabolic risk in Indian children, adolescents and youth with type 1 diabetes mellitus. *Endocrine* **2021**, *5*. [CrossRef] [PubMed]
30. Kilpatrick, E.S.; Rigby, A.S.; Atkin, S.L. Insulin resistance, the metabolic syndrome, and complication risk in type 1 diabetes: "double diabetes" in the Diabetes Control and Complications Trial. *Diabetes Care* **2007**, *30*, 707–712. [CrossRef]
31. Reinehr, T.; Holl, R.W.; Roth, C.L.; Wiesel, T.; Stachow, R.; Wabitsch, M.; Andler, W.; DPV-Wiss Study Group. Insulin resistance in children and adolescents with type 1 diabetes mellitus: Relation to obesity. *Pediatr. Diabetes* **2005**, *6*, 5–12. [CrossRef]
32. Fröhlich-Reiterer, E.E.; Rosenbauer, J.; Bechtold-Dalla Pozza, S.; Hofer, S.E.; Schober, E.; Holl, R.W.; DPV-Wiss Study Group; German BMBF Competence Networks Diabetes Mellitus and Obesity. Predictors of increasing BMI during the course of diabetes in children and adolescents with type 1 diabetes: Data from the German/Austrian DPV multicentre survey. *Arch. Dis. Child.* **2014**, *99*, 738–743. [CrossRef]
33. da Costa, V.M.; de Carvalho Padilha, P.; de Lima, G.C.; Ferreira, A.A.; Luescher, J.L.; Porto, L.; Peres, W.A. Overweight among children and adolescent with type I diabetes mellitus: Prevalence and associated factors. *Diabetol. Metab. Syndr.* **2016**, *8*, 39. [CrossRef]
34. Mochizuki, M.; Ito, Y.; Yokomichi, H.; Kikuchi, T.; Soneda, S.; Musha, I.; Anzou, M.; Kobayashi, K.; Matsuo, K.; Sugihara, S.; et al. Increasing secular trends in height and obesity in children with type 1 diabetes: JSGIT cohort. *PLoS ONE* **2020**, *15*, e0242259. [CrossRef]

35. Gilbertson, H.R.; Reed, K.; Clark, S.; Francis, K.L.; Cameron, F.J. An audit of the dietary intake of Australian children with type 1 diabetes. *Nutr. Diabetes* **2018**, *8*, 10. [CrossRef]
36. Van der Schueren, B.; Ellis, D.; Faradji, R.N.; Al-Ozairi, E.; Rosen, J.; Mathieu, C. Obesity in people living with type 1 diabetes. *Lancet Diabetes Endocrinol.* **2021**, *9*, 776–785. [CrossRef]
37. March, C.A.; Becker, D.J.; Libman, I.M. Nutrition and Obesity in the Pathogenesis of Youth-Onset Type 1 Diabetes and Its Complications. *Front. Endocrinol.* **2021**, *12*, 622901. [CrossRef] [PubMed]
38. Mehta, S.N.; Haynie, D.L.; Higgins, L.A.; Bucey, N.N.; Rovner, A.J.; Volkening, L.K.; Nansel, T.R.; Laffel, L.M. Emphasis on carbohydrates may negatively influence dietary patterns in youth with type 1 diabetes. *Diabetes Care* **2009**, *32*, 2174–2176. [CrossRef] [PubMed]
39. Nansel, T.R.; Haynie, D.L.; Lipsky, L.M.; Laffel, L.M.; Mehta, S.N. Multiple indicators of poor diet quality in children and adolescents with type 1 diabetes are associated with higher body mass index percentile but not glycemic control. *J. Acad. Nutr. Diet.* **2012**, *112*, 1728–1735. [CrossRef] [PubMed]
40. Monzani, A.; Ricotti, R.; Caputo, M.; Solito, A.; Archero, F.; Bellone, S.; Prodam, F. A Systematic Review of the Association of Skipping Breakfast with Weight and Cardiometabolic Risk Factors in Children and Adolescents. What Should We Better Investigate in the Future? *Nutrients* **2019**, *11*, 387. [CrossRef] [PubMed]
41. Ricotti, R.; Caputo, M.; Monzani, A.; Pigni, S.; Antoniotti, V.; Bellone, S.; Prodam, F. Breakfast Skipping, Weight, Cardiometabolic Risk, and Nutrition Quality in Children and Adolescents: A Systematic Review of Randomized Controlled and Intervention Longitudinal Trials. *Nutrients* **2021**, *13*, 3331. [CrossRef] [PubMed]
42. Ricotti, R.; Caputo, M.; Prodam, F. Chapter 9. In *The Mediterranean Diet, An Evidence-Based Approach*, 2nd ed.; Preedy, V.R., Watson, R.R., Eds.; Elsevier: London, UK, 2020; pp. 89–93. ISBN 978-0-12-818649-7.
43. Jacob, R.; Bertrand, C.; Llewellyn, C.; Couture, C.; Labonté, M.È.; Tremblay, A.; Bouchard, C.; Drapeau, V.; Pérusse, L. Dietary Mediators of the Genetic Susceptibility to Obesity—Results from the Quebec Family Study. *J. Nutr.* **2021**, *152*, 49–58. [CrossRef] [PubMed]
44. Calleja, M.; Caetano Feitoza, N.; Falk, B.; Klentrou, P.; Ward, W.E.; Sullivan, P.J.; Josse, A.R. Increased dairy product consumption as part of a diet and exercise weight management program improves body composition in adolescent females with overweight and obesity—A randomized controlled trial. *Pediatr. Obes.* **2020**, *15*, e12690. [CrossRef]
45. Dougkas, A.; Barr, S.; Reddy, S.; Summerbell, C.D. A critical review of the role of milk and other dairy products in the development of obesity in children and adolescents. *Nutr. Res. Rev.* **2019**, *32*, 106–127. [CrossRef]
46. Wang, W.; Wu, Y.; Zhang, D. Association of dairy products consumption with risk of obesity in children and adults: A meta-analysis of mainly cross-sectional studies. *Ann. Epidemiol.* **2016**, *26*, 870–882.e2. [CrossRef]
47. Lu, L.; Xun, P.; Wan, Y.; He, K.; Cai, W. Long-term association between dairy consumption and risk of childhood obesity: A systematic review and meta-analysis of prospective cohort studies. *Eur. J. Clin. Nutr.* **2016**, *70*, 414–423. [CrossRef] [PubMed]
48. Ağagündüz, D.; Yılmaz, B.; Şahin, T.Ö.; Güneşliol, B.E.; Ayten, Ş.; Russo, P.; Spano, G.; Rocha, J.M.; Bartkiene, E.; Özogul, F. Dairy Lactic Acid Bacteria and Their Potential Function in Dietetics: The Food-Gut-Health Axis. *Foods* **2021**, *10*, 3099. [CrossRef] [PubMed]
49. Willett, W.C.; Ludwig, D.S. Milk and Health. *N. Engl. J. Med.* **2020**, *382*, 644–654. [CrossRef] [PubMed]
50. Ahola, A.J.; Forsblom, C.M.; Harjutsalo, V.; Groop, P.H. Nut Consumption Is Associated with Lower Risk of Metabolic Syndrome and Its Components in Type 1 Diabetes. *Nutrients* **2021**, *13*, 3909. [CrossRef] [PubMed]
51. Toeller, M.; Buyken, A.E.; Heitkamp, G.; Cathelineau, G.; Ferriss, B.; Michel, G.; EURODIAB IDDM Complications Study Group. Nutrient intakes as predictors of body weight in European people with type 1 diabetes. *Int. J. Obes. Relat. Metab. Disord.* **2001**, *25*, 1815–1822. [CrossRef] [PubMed]
52. Fernández-Rodríguez, R.; Mesas, A.E.; Garrido-Miguel, M.; Martínez-Ortega, I.A.; Jiménez-López, E.; Martínez-Vizcaíno, V. The Relationship of Tree Nuts and Peanuts with Adiposity Parameters: A Systematic Review and Network Meta-Analysis. *Nutrients* **2021**, *13*, 2251. [CrossRef]
53. Guarneiri, L.L.; Cooper, J.A. Intake of Nuts or Nut Products Does Not Lead to Weight Gain, Independent of Dietary Substitution Instructions: A Systematic Review and Meta-Analysis of Randomized Trials. *Adv. Nutr. Int. Rev. J.* **2021**, *12*, 384–401. [CrossRef]
54. Liu, X.; Li, Y.; Guasch-Ferré, M.; Willett, W.C.; Drouin-Chartier, J.P.; Bhupathiraju, S.N.; Tobias, D.K. Changes in nut consumption influence long-term weight change in US men and women. *BMJ Nutr. Prev. Health* **2019**, *2*, 90–99. [CrossRef]
55. Ros, E.; Singh, A.; O'Keefe, J.H. Nuts: Natural Pleiotropic Nutraceuticals. *Nutrients* **2021**, *13*, 3269. [CrossRef]
56. Fortins, R.F.; Lacerda, E.M.A.; Silverio, R.N.C.; do Carmo, C.N.; Ferreira, A.A.; Felizardo, C.; do Nascimento, B.F.; Luescher, J.L.; Padilha, P.C. Predictor factors of glycemic control in children and adolescents with type 1 diabetes mellitus treated at a referral service in Rio de Janeiro, Brazil. *Diabetes Res. Clin. Pract.* **2019**, *154*, 138–145. [CrossRef]
57. Wolever, T.M.; Hamad, S.; Chiasson, J.L.; Josse, R.G.; Leiter, L.A.; Rodger, N.W.; Ross, S.A.; Ryan, E.A. Day-to-day consistency in amount and source of carbohydrate intake associated with improved blood glucose control in type 1 diabetes. *J. Am. Coll. Nutr.* **1999**, *18*, 242–247. [CrossRef] [PubMed]
58. Jebraeili, H.; Shabbidar, S.; Sajjadpour, Z.; Aghdam, S.D.; Qorbani, M.; Rajab, A.; Sotoudeh, G. The association between carbohydrate quality index and anthropometry, blood glucose, lipid profile and blood pressure in people with type 1 diabetes mellitus: A cross-sectional study in Iran. *J. Diabetes Metab. Disord.* **2021**, *20*, 1349–1358. [CrossRef] [PubMed]

59. Chiavaroli, L.; Lee, D.; Ahmed, A.; Cheung, A.; Khan, T.A.; Blanco, S.; Mejia, S.B.; Mirrahimi, A.; Jenkins, D.J.A.; Livesey, G.; et al. Effect of low glycaemic index or load dietary patterns on glycaemic control and cardiometabolic risk factors in diabetes: Systematic review and meta-analysis of randomised controlled trials. *BMJ* **2021**, *374*, 1651. [CrossRef] [PubMed]
60. Jacobsen, S.S.; Vistisen, D.; Vilsbøll, T.; Bruun, J.M.; Ewers, B. The quality of dietary carbohydrate and fat is associated with better metabolic control in persons with type 1 and type 2 diabetes. *Nutr. J.* **2020**, *19*, 125. [CrossRef]
61. Cherubini, V.; Marino, M.; Marigliano, M.; Maffeis, C.; Zanfardino, A.; Rabbone, I.; Giorda, S.; Schiaffini, R.; Lorubbio, A.; Rollato, S.; et al. Rethinking Carbohydrate Intake and Time in Range in Children and Adolescents with Type 1 Diabetes. *Nutrients* **2021**, *13*, 3869. [CrossRef]
62. Möllsten, A.V.; Dahlquist, G.G.; Stattin, E.L.; Rudberg, S. Higher intakes of fish protein are related to a lower risk of microalbuminuria in young Swedish type 1 diabetic patients. *Diabetes Care* **2001**, *24*, 805–810. [CrossRef]
63. Mendivil, C.O. Fish Consumption: A Review of Its Effects on Metabolic and Hormonal Health. *Nutr. Metab. Insights* **2021**, *14*, 1–6. [CrossRef]
64. Paterson, M.A.; King, B.R.; Smart, C.E.M.; Smith, T.; Rafferty, J.; Lopez, P.E. Impact of dietary protein on postprandial glycaemic control and insulin requirements in Type 1 diabetes: A systematic review. *Diabet. Med.* **2019**, *36*, 1585–1599. [CrossRef]
65. Furthner, D.; Lukas, A.; Schneider, A.M.; Mörwald, K.; Maruszczak, K.; Gombos, P.; Gomahr, J.; Steigleder-Schweiger, C.; Weghuber, D.; Pixner, T. The Role of Protein and Fat Intake on Insulin Therapy in Glycaemic Control of Paediatric Type 1 Diabetes: A Systematic Review and Research Gaps. *Nutrients* **2021**, *13*, 3558. [CrossRef]
66. Purdel, C.; Ungurianu, A.; Margina, D. Metabolic and Metabolomic Insights Regarding the Omega-3 PUFAs Intake in Type 1 Diabetes Mellitus. *Front. Mol. Biosci.* **2021**, *8*, 783065. [CrossRef]
67. Cicero, A.F.G.; Veronesi, M.; Fogacci, F. Dietary Intervention to Improve Blood Pressure Control: Beyond Salt Restriction. *High Blood Press. Cardiovasc. Prev.* **2021**, *28*, 547–553. [CrossRef] [PubMed]
68. De Pergola, G.; D'Alessandro, A. Influence of Mediterranean Diet on Blood Pressure. *Nutrients* **2018**, *10*, 1700. [CrossRef] [PubMed]
69. Günther, A.L.; Liese, A.D.; Bell, R.A.; Dabelea, D.; Lawrence, J.M.; Rodriguez, B.L.; Standiford, D.A.; Mayer-Davis, E.J. Association between the dietary approaches to hypertension diet and hypertension in youth with diabetes mellitus. *Hypertension* **2009**, *53*, 6–12. [CrossRef] [PubMed]
70. Lalji, R.; Tullus, K. What's new in paediatric hypertension? *Arch. Dis. Child.* **2018**, *103*, 96–100. [CrossRef] [PubMed]
71. Cohen, L.; Curhan, G.; Forman, J. Association of Sweetened Beverage Intake with Incident Hypertension. *J. Gen. Intern. Med.* **2012**, *27*, 1127–1134. [CrossRef] [PubMed]
72. Appel, L.J.; Sacks, F.M.; Carey, V.J.; Obarzanek, E.; Swain, J.F.; Miller, E.R., III; Conlin, P.R.; Erlinger, T.P.; Rosner, B.A.; Laranjo, N.M.; et al. OmniHeart Collaborative Research Group. Effects of protein, monounsaturated fat, and carbohydrate intake on blood pressure and serum lipids: Results of the OmniHeart randomized trial. *JAMA* **2005**, *294*, 2455–2464. [CrossRef]
73. Godos, J.; Tieri, M.; Ghelfi, F.; Titta, L.; Marventano, S.; Lafranconi, A.; Gambera, A.; Alonzo, E.; Sciacca, S.; Buscemi, S.; et al. Dairy foods and health: An umbrella review of observational studies. *Int. J. Food Sci. Nutr.* **2020**, *71*, 138–151. [CrossRef]
74. Katsiki, N.; Pérez-Martínez, P.; Lopez-Miranda, J. Olive Oil Intake and Cardiovascular Disease Prevention: "Seek and You Shall Find". *Curr. Cardiol. Rep.* **2021**, *23*, 64. [CrossRef]
75. Massaro, M.; Scoditti, E.; Carluccio, M.A.; Calabriso, N.; Santarpino, G.; Verri, T.; De Caterina, R. Effects of Olive Oil on Blood Pressure: Epidemiological, Clinical, and Mechanistic Evidence. *Nutrients* **2020**, *12*, 1548. [CrossRef]
76. Cutruzzolà, A.; Parise, M.; Vallelunga, R.; Lamanna, F.; Gnasso, A.; Irace, C. Effect of Extra Virgin Olive Oil and Butter on Endothelial Function in Type 1 Diabetes. *Nutrients* **2021**, *13*, 2436. [CrossRef]
77. Cameron, A.J.; Spence, A.C.; Laws, R.; Hesketh, K.D.; Lioret, S.; Campbell, K.J. A Review of the Relationship Between Socioeconomic Position and the Early-Life Predictors of Obesity. *Curr. Obes. Rep.* **2015**, *4*, 350–362. [CrossRef] [PubMed]
78. Patro, B.; Liber, A.; Zalewski, B.; Poston, L.; Szajewska, H.; Koletzko, B. Maternal and paternal body mass index and offspring obesity: A systematic review. *Ann. Nutr. Metab.* **2013**, *63*, 32–41. [CrossRef] [PubMed]
79. Nielsen, L.A.; Nielsen, T.R.; Holm, J.C. The Impact of Familial Predisposition to Obesity and Cardiovascular Disease on Childhood Obesity. *Obes. Facts* **2015**, *8*, 319–328. [CrossRef] [PubMed]
80. Mazarello Paes, V.; Charalampopoulos, D.; Edge, J.; Taylor-Robinson, D.; Stephenson, T.; Amin, R. Predictors of glycemic control in the first year of diagnosis of childhood onset type 1 diabetes: A systematic review of quantitative evidence. *Pediatr. Diabetes* **2018**, *19*, 18–26. [CrossRef]
81. Hannonen, R.; Aunola, K.; Eklund, K.; Ahonen, T. Maternal Parenting Styles and Glycemic Control in Children with Type 1 Diabetes. *Int. J. Environ. Res. Public Health* **2019**, *16*, 214. [CrossRef] [PubMed]
82. Mahfouz, E.M.; Kamal, N.N.; Mohammed, E.S.; Refaei, S.A. Effects of Mothers' Knowledge and Coping Strategies on the Glycemic Control of Their Diabetic Children in Egypt. *Int. J. Prev. Med.* **2018**, *9*, 26. [CrossRef] [PubMed]
83. Wu, N.; Bredin, S.S.D.; Jamnik, V.K.; Koehle, M.S.; Guan, Y.; Shellington, E.M.; Li, Y.; Li, J.; Warburton, D.E.R. Association between physical activity level and cardiovascular risk factors in adolescents living with type 1 diabetes mellitus: A cross-sectional study. *Cardiovasc. Diabetol.* **2021**, *20*, 62. [CrossRef] [PubMed]

MDPI

Article

Adherence to a Mediterranean-Style Dietary Pattern and Cancer Risk in a Prospective Cohort Study

Ioanna Yiannakou [1,2], Martha R. Singer [1], Paul F. Jacques [3], Vanessa Xanthakis [1,4], R. Curtis Ellison [1] and Lynn L. Moore [1,2,*]

1 Department of Medicine, Preventive Medicine and Epidemiology, Boston University School of Medicine, Boston, MA 02118, USA; ioannay@bu.edu (I.Y.); msinger@bu.edu (M.R.S.); Vanessax@bu.edu (V.X.); ellison@bu.edu (R.C.E.)
2 Graduate Programs in Nutrition and Metabolism, Boston University School of Medicine, Boston, MA 02118, USA
3 Nutritional Epidemiology, Jean Mayer USDA Human Nutrition Research Center on Aging at Tufts University, Boston, MA 02111, USA; paul.jacques@tufts.edu
4 Department of Biostatistics, Boston University School of Public Health, Boston, MA 02118, USA
* Correspondence: llmoore@bu.edu; Tel.: +1-617-358-1325

Abstract: A Mediterranean-style diet is a healthy eating pattern that may benefit cancer risk, but evidence among Americans is scarce. We examined the prospective association between adherence to such a diet pattern and total cancer risk. A Mediterranean-style dietary pattern (MSDP) score was derived from a semi-quantitative food frequency questionnaire at exam 5 (1991–1995). Subjects included 2966 participants of the Framingham Offspring Study who were free of prevalent cancer. Cox proportional hazards regression models were used to estimate hazard ratios (HRs) and 95% confidence intervals (CIs), adjusting for demographic, lifestyle, and anthropometric measures. Cox-models were also used to examine effect modification by lifestyle and anthropometric measures. During 18 years of median follow-up, 259 women and 352 men were diagnosed with cancer. Women with moderate or higher adherence to the MSDP had ≥25% lower risks of cancer than women with the lowest MSDP (HR (moderate vs. lowest): 0.71, 95% CI: 0.52–0.97 and HR (highest vs. lowest): 0.74; 95% CI: 0.55–0.99). The association between MSDP score and cancer risk in men was weaker except in non-smokers. Beneficial effects of the MSDP in women were stronger among those who were not overweight. In this study, higher adherence to MSDP was associated with lower cancer risk, especially among women.

Keywords: cancer; Mediterranean diet; diet patterns; cohort study; epidemiology

Citation: Yiannakou, I.; Singer, M.R.; Jacques, P.F.; Xanthakis, V.; Ellison, R.C.; Moore, L.L. Adherence to a Mediterranean-Style Dietary Pattern and Cancer Risk in a Prospective Cohort Study. *Nutrients* **2021**, *13*, 4064. https://doi.org/10.3390/nu13114064

Academic Editor: Daniela Bonofiglio

Received: 26 October 2021
Accepted: 12 November 2021
Published: 13 November 2021

1. Introduction

The American Institute for Cancer Research (2018) report [1] recommends a healthy dietary pattern rich in fruits, vegetables, legumes, and whole grains while limiting consumption of added sugars and red and processed meats. Many of these recommendations are consistent with a Mediterranean-style diet, which has been suggested in the United States (US), starting with the 2015–2020 Dietary Guidelines, as a healthy eating pattern [2]. The Mediterranean diet has been long described as a well-balanced diet with a predominance of plant-based food sources. However, the specific foods consumed are somewhat variable across different Mediterranean cultures [3–5]. The diet in Crete prior to1960 is often considered the model for a traditional Mediterranean diet. It is characterized by higher intakes of vegetables, fruits, legumes, nuts, and non–refined cereals and grain products. Fish and poultry are consumed in moderation, whereas red and processed meats, dairy products, refined grains, and sweets are limited. Olive oil and olives are the most common sources of fat. Red wine is consumed in moderation during meals, and the population is generally quite physically active [5,6].

For nearly three decades, epidemiologic evidence has supported the health benefits of adherence to a Mediterranean diet in the primary and secondary prevention of non–communicable chronic diseases [3,7,8]. Although most cohort studies suggest a protective association between higher adherence to a Mediterranean-style diet and risk of specific cancers [9–12], evidence of its effects on total cancer risk is very limited. Because a Mediterranean-style diet has much in common with an anti-inflammatory diet, it is possible that the diet pattern may have more generalized beneficial effects on cancer risk [13]. Two previous prospective analyses were conducted within the European Prospective Investigation into Cancer and Nutrition cohort (EPIC), but the results were somewhat inconsistent [14,15]. In the first, higher adherence to a Mediterranean diet was strongly associated with a lower cancer risk among Greek women and a lower overall risk of non-smoking-related cancers in both men and women [14]. In the second study, conformity to a Mediterranean diet was protective against cancer occurrence in both Mediterranean and non-Mediterranean countries, although these results were weaker than those in the Greek cohort, and no sex-specific differences were observed [15].

Mediterranean diet studies sometimes differ in the means by which the diet pattern is scored. Many studies, including EPIC, have used a Mediterranean diet score based on whether the participant had higher or lower intakes of the relevant foods and nutrients, defined as being above or below the sex-median intakes in that population [16]. Because the actual intakes of these Mediterranean-style foods and nutrients differ widely between population groups, the scores may not uniformly reflect higher adherence to a traditional Mediterranean-style diet as defined by the diet pyramid. A new Mediterranean-style dietary pattern (MSDP) score that does not rely on median intakes was developed in 2009 using data from the Framingham Study [17]. We will use this score to assess adherence to the MSDP.

The primary aim of this prospective study was to examine the longitudinal association between adherence to the MSDP and total cancer risk in the Framingham Offspring Study (FOS) cohort. We examined whether this association was modified by anthropometric measures of body fat or lifestyle factors. In secondary analyses, we assessed the association between each food group considered in the MSDP and total cancer risk.

2. Materials and Methods

2.1. Study Population

In 1971, 5124 individuals were enrolled in the prospective FOS. The participants were the children of those who were part of the original Framingham Heart Study. The examination visits, starting with exam 2, were carried out at approximately 4-year intervals [17]. Food frequency questionnaires were used to assess diet starting at exam 5, the baseline visit for these analyses (1991–1995). Participants were followed for the development of cancer until 2013. The final study sample included a total of 2966 individuals, as shown in Supplementary Figure S1. A total of 3712 participants attended examination visit 5. Of these, we excluded the following participants: (a) missing or invalid FFQ data ($n = 362$); (b) history of cancer or prevalent cancer except non-melanoma skin cancer ($n = 147$); (c) aged less than 30 years at baseline ($n = 4$); and (d) missing covariates ($n = 233$). The Framingham Offspring Study data collection and these analyses were approved by the Institutional Review Board of Boston University School of Medicine (Protocols H-32086 and H-32132).

2.2. Dietary Assessment and Adherence to the MSDP

Diet data was assessed by self-report using a semi–quantitative food frequency questionnaire (FFQ) [18]. The FFQ includes a list of 126 food items and assesses frequency of consumption of each food during the previous year, with responses ranging from "never or <1 servings/month" to "≥6 servings/day." Separate questions queried types of breakfast cereals and cooking oils consumed.

The scoring methods for the MSDP score were based on adherence to a traditional Mediterranean diet [19] and have been previously described [17]. It includes data on the recommended intake of the following 13 food groups in the Mediterranean diet pyramid: wholegrain cereals, fruit, vegetables, dairy, wine, fish, poultry, olives/legumes/nuts, potatoes, eggs, sweets, meat, and olive oil. Foods from an American diet pattern that were similar to those in a traditional Mediterranean diet were also included. For example, yams are not part of the potato group in a traditional Mediterranean diet but were included in the potato category in the MSDP. The use of olive oil was scored as follows: (a) used exclusively (score = 10), (b) used olive oil and other vegetable oils (score = 5), or (c) used no olive oil (score = 0). All other foods were scored from 0 to 10, based on percent adherence to Mediterranean Diet Pyramid recommendations (e.g., consuming 80% of the recommended amount for a food category yielded a score of 8). This MSDP includes a penalty for overconsumption as well as underconsumption (e.g., exceeding the maximum recommended intake by 30% would result in a score of 7). The maximum penalty was 10 points. The total of the 13 component scores was standardized to a scale of 0–100 and weighted (from 0 to 1) by the percent of total energy derived from consuming foods included in the Mediterranean diet pyramid. For example, if 45% of energy was derived from foods not included on the Mediterranean diet pyramid, the calculated weight was 0.55. The final MSDP score ranged from 0 to 100.

2.3. Cancer Outcomes

The primary occurrence of cancer was adjudicated using a standardized Framingham Study protocol as previously described [20]. Briefly, possible cancer cases were initially detected using self–report at each examination visit, surveillance of local hospital admissions, and searches of death records from the state health department and the National Death Index [21]. Framingham investigators gathered pathology reports and other clinical and laboratory data for each subject [22] for the purpose of establishing the correct cancer diagnosis and date of diagnosis [23]. In these analyses, there were 611 cancer cases identified as first primary malignant cancers; the topography and morphologic characteristics of each cancer were coded based on the International Classification of Diseases for Oncology code (ICD-O-3 only). Non–melanoma skin cancer cases were excluded in these analyses.

2.4. Potential Confounders and Effect Modifiers

Height and weight were measured with a standard beam balance scale with the subject wearing a hospital gown and no shoes [23]. To reduce random error and the effect of natural height loss occurring after the age of 60, we calculated each participant's average height from all exam visits up until the age of 60 years [24]. We used this average height in combination with baseline weight at exam 5 to calculate baseline body mass index (BMI) (i.e., weight (kg) at exam 5 divided by mean height (m^2).

Physical activity was assessed at exams 4 (1987–1991) and 7 (1998–2001) using self-report questionnaires that included recreational activities and household work. The intensity levels for light, moderate, and vigorous activities were derived from previous studies of oxygen utilization for a given level of activity. A weighted moderate and vigorous activity score was calculated by summing total hours of moderate activity (multiplied by its intensity value) and total hours of vigorous activity (multiplied by its intensity score) [25].

The number of years of education was self-reported at exam two and was used to classify baseline education level into three categories: high school or less, some college, and college or graduate degree. Missing education data at exam two were imputed hierarchically as follows: education level at exam 8 (2005–2008), median years of education level for the subjects with the same occupation at exam 7 (1998–2001), and sex-specific median years of education at exam 2 (1979–1983). Missing data for self-reported pack-years of cigarette smoking at exam 5 (1991–1995) were substituted with the mean of pack-years from exams 4 (1987–1991) and 6 (1995–1998) when available. Other self-reported covariates at exam 5 (1991–1995) included: multivitamin use, other supplement use, cigarette smoking status (never, former, or current smoking defined as $\geq$1 cigarettes per day), energy intake, and alcohol intake (g/day). Measures of abdominal adiposity, including waist circumference (cm) and waist–to–height ratio (missing values substituted using the mean from exams 4 and 6), were also collected at baseline. Further, we created a time-dependent variable to reflect self-reported estrogen use (including estrogen use only). We classified women as never or ever users of estrogen based on self-report across multiple exam visits. Type 2 diabetes was defined at baseline when the participant met one of the following criteria: (a) 10-hour-fasting glucose of ≥126 mg/dL; (b) non-fasting glucose of ≥200 mg/dL; (c) confirmed treatment of diabetes, or (d) self-reported diagnosis of possible diabetes at one visit with a subsequent diagnosis of definite diabetes at the next exam (in the absence of an excessive weight gain of 7% or more of body weight).

2.5. Statistical Analysis

Sensitivity analyses were based on the exposure distribution and power considerations and used to explore different cut-off values for categorizing the MSDP score. The optimal classification was selected as follows: low: 4.0–19.0 (reference group), moderate: 19.1–25.0, and high: 25.1–50.9. Incidence rates for total cancer were computed by dividing the number of cancer cases by total person-years (py) of follow-up calculated from baseline (exam 5, 1991–1995) to the first of the following events: primary occurrence of cancer, loss to follow-up, date of last exam or death. Age and multivariable-adjusted Cox proportional hazards regression models were used to estimate hazard ratios (HRs) and 95% confidence intervals (CIs) for incident cancer. A test for linear trend across MSDP score categories was performed.

Because the effects of adherence to the MSDP may depend on the level of other lifestyle and anthropometric measures, we chose to assess interaction between MSDP adherence and each of the following risk factors: BMI, WHtR, alcohol intake, and cigarette smoking status. To better assess public health importance, interaction was assessed on an additive rather than a multiplicative scale [26,27]. Cox-models were used to estimate the relative excess risk due to additive interaction in men and women separately [28]. To optimize statistical power for these analyses, we used sensitivity analysis to dichotomize the MSDP score as ≤19 (low) vs. >19 (moderate/high). For these analyses, BMI was categorized as <25 vs. ≥25 kg/m^2 for women and <30 vs. ≥30 kg/m^2 for men [21,23]. The WHtR was classified as <57 vs. ≥57 for women and<51 vs. ≥51 cm/m for men [22]. Three categories of current alcohol intake were chosen to represent non-drinkers, light-to-moderate drinkers, and heavy drinkers: 0 g/day, 0.1–13.99 g/day, and ≥14.00 g/day, respectively, for men and 0 g/day, 0.1–6.9 g/day, and ≥7 g/day, respectively, for women. Cigarette smoking status was classified as current smokers, former smokers, and non-smokers.

The final multivariable models adjusted for confounders that were found to alter the age-adjusted hazard ratios by approximately 10% or more in men and women separately. Those factors retained in the final models included BMI (kg/m^2), cigarette pack-years, physical activity (metabolic equivalent hours/day), prevalent diabetes, and supplement use (never vs. ever). Factors not included in the final models were educational level, alcohol intake, and estrogen use because they were not found confounding the association between MSDP adherence and total cancer risk.

Finally, we analyzed the association between each food group considered in the MSDP score and total cancer risk. Due to substantially right-skewed data, we excluded those with the highest 1% of intake for each food group and then used sensitivity analyses to determine the cut-off values to define intake as low (reference), moderate, or high for each food group. The multivariable Cox models included the same factors listed above plus total energy intake; these models also mutually adjusted for the intakes of all other MSDP score components. A test for linear trend across each food group was performed based on the category-specific medians of food intake. No violations of the proportional hazard assumptions were found in any of the models. The statistical analyses in this study were conducted using SAS statistical software, version 9.4 (SAS Institute, Cary, NC, USA).

3. Results

The MSDP score at baseline was normally distributed, with a mean of 22.4 (± 7.3) (range 4.0–51.0). Table 1 shows the subject characteristics according to MSDP score categories (low, moderate, high). Compared with participants in the lowest category, those with the highest scores were older, more likely to be women, had higher educational levels, slightly lower alcohol intakes, and more often used dietary supplements. In addition, they were also less likely to be current smokers. Women in the highest MSDP score category were more likely to be estrogen users.

Table 1. Baseline characteristics according to MSDP score categories in the Framingham Offspring Study.

	MSDP Score Categories		
Characteristic (*n* = 2966)	**Low (4.0–19.0)** *n* = **995**	**Moderate (19.1–25.0)** *n* = **951**	**High (25.1–50.9)** *n* = **1020**
Sex, *n* (%)			
Women	443 (45)	505 (53)	630 (62)
Men	592 (56)	446 (47)	390 (38)
Age (years)	52.8 (± 9.5)	55.0 (± 9.9)	55.2 (± 9.3)
Age at diagnosis (years)	70.3 (± 8.9)	72.5 (± 9.4)	73.0 (± 8.9)
Education, *n* (%)			
≤High School	445 (44.7)	351 (36.9)	326 (32.0)
Some college	265 (26.6)	283 (29.8)	316 (31.0)
College, Graduate degree	285 (28.6)	317 (33.3)	378 (37.1)
BMI (kg/m^2)	27.4 (± 5.0)	27.4 (± 4.8)	27.1 (± 5.0)
Waist (cm)	93.8 (± 14.2)	93.0 (± 14.1)	91.1 (± 14.3)
Waist-to-height ratio	0.55 (± 0.08)	0.55 (± 0.08)	0.55 (± 0.08)
Cigarette smoking, *n* (%)			
Never	311 (31.3)	325 (34.3)	416 (40.8)
Former	406 (40.8)	450 (47.4)	478 (46.9)
Current	378 (27.9)	174 (18.3)	126 (12.3)
Pack years of smoking [1]	20.0 (± 0.9)	16.1 (± 0.8)	12.0 (± 0.6)
Energy intake (kcals/day)	1741 (± 650)	1885 (± 611)	1963 (± 580)
Alcohol (g/day) [1]	8.0 (± 0.4)	7.3 (± 0.3)	6.7 (± 0.3)
Supplement use, *n* (%)			
No	735 (73.9)	680 (71.5)	636 (62.4)
Yes	242 (24.3)	250 (26.3)	367 (36.0)
Physical activity (METs/day)	14.3 (± 9.2)	14.5 (± 8.1)	15.1 (± 8.0)
Prevalent diabetes, *n* (%)	59 (5.9)	65 (6.8)	79 (7.8)
Estrogen use, *n* (%) [2]			
Never	379 (85.8)	426 (84.4)	507 (80.1)
Ever	63 (14.2)	79 (15.6)	122 (19.4)

Values are expressed as mean (±SD) or n (column percentage) or otherwise stated. [1] Values are expressed as geometric mean ± SE. [2] Sample includes 1576 women as 2 were missing estrogen use data. MSDP, Mediterranean-style dietary pattern; BMI, body mass index; D, day; and METs, Metabolic equivalents.

Table 2 shows the median intakes (servings per day or per week) for each of the 13 food groups considered in the MSDP score along with their MSDP scores. Median scores were highest for poultry, potatoes, dairy, and fruits. The lowest scores were for the categories of sweets, meat, olive oil use, and wine consumption.

Table 2. Intake and score distributions of the food groups considered in the MSDP score among the participants in the Framingham Offspring Study.

Food Groups Considered in the MSDP Score	Median Intakes (5%–95%)	Median Scores (5%–95%) [1]
	Servings/day	
Whole grains	0.90 (0.00–3.60)	1.13 (0.00–4.50)
Fruit	1.50 (0.20–4.40)	4.67 (0.33–9.33)
Vegetables	2.40 (0.80–5.80)	4.00 (1.17–8.50)
Dairy	1.20 (0.20–4.10)	5.00 (0.00–9.00)
Wine: Men	0.10 (0.00–0.90)	0.33 (0.00–3.00)
Women	0.10 (0.00–0.90)	0.67 (0.00–5.33)
	Servings/week	
Olives, pulses, and nuts	1.20 (0.00–5.10)	3.00 (0.00–9.25)
Potatoes	3.50 (0.50–7.60)	5.00 (0.00–10.00)
Poultry	5.00 (0.80–9.90)	5.75 (0.00–8.25)
Eggs	0.50 (0.00–3.00)	1.67 (0.00–10.00)
Meat	4.00 (0.50–11.90)	0.00 (0.00–8.00)
Sweets	13.80 (1.90–45.00)	0.00 (0.00–7.67)
Fish	2.30 (0.50–7.00)	3.73 (0.67–9.00)
Olive oil use score [2]		0.00 (0.00–10.00)

[1] Scores ranged from 0 to 10 based on percent adherence to Mediterranean Diet Pyramid recommendations, except for the use of olive oil. [2] Olive oil was scored as (a) used exclusively (score = 10), (b) used olive oil and other vegetable oils (score = 5), or (c) used no olive oil (score = 0). MSDP, Mediterranean style dietary pattern.

Of the 2966 men and women at baseline, 611 subsequently developed cancer (Table 3). Overall, those with higher MSDP scores had lower cancer incidence rates (1160 cases/10,000 py of follow-up) compared to those with moderate and lower MSDP scores, 1240 and 1416 cases/10,000 py, respectively. In the fully adjusted models, participants with moderate and higher MSDP scores had non-statistically significant 15% and 16% lower risks of total cancer than those in the lowest MSDP score category, respectively ($HR_{moderate}$: 0.85, 95% CI: 0.70–1.04; HR_{high}: 0.84, 95% CI: 0.68–1.03). In sex-stratified analyses, greater adherence to a Mediterranean diet had a much stronger beneficial effect on total cancer risk in women than in men. For women, those in the moderate and higher MSDP score categories had 29% (95% CI: 0.52–0.97) and 26% (95% CI: 0.55–0.99) lower cancer risks than those in the lowest MSDP score category.

Table 3. Hazard ratios for total cancer, according to MSDP score categories in the Framingham Offspring Study.

	Subjects	Cases/PY	Rate/10,000 py	HR (95% CI) [1]	HR (95% CI) [2]
All subjects (*n* = 2966)					
MSDP score					
Low (4.0–19.0)	995	226/15,964	1416	1.00 (Ref.)	1.00 (Ref.)
Moderate (19.1–25.0)	951	191/15,407	1240	0.82 (0.67–0.99)	0.85 (0.70–1.04)
High (25.1–50.9)	1020	194/16,717	1160	0.79 (0.65–0.96)	0.84 (0.69–1.03)
P-trend				0.02	0.09
Women (*n* = 1578)					
MSDP score					
Low (4.0–19.0)	443	88/7333	1200	1.00 (Ref.)	1.00 (Ref.)
Moderate (19.1–25.0)	505	74/8552	865	0.86 (0.50–0.93)	0.71 (0.52–0.97)
High (25.1–50.9)	630	97/10,718	905	0.70 (0.53–0.94)	0.74 (0.55–0.99)
P-trend				0.02	0.05
Men (*n* = 1388)					
MSDP score					
Low (4.0–19.0)	552	138/8632	1599	1.00 (Ref.)	1.00 (Ref.)
Moderate (19.1–25.0)	446	117/6856	1707	0.90 (0.70–1.16)	0.95 (0.74–1.22)
High (25.1–47.7)	390	97/5999	1617	0.83 (0.64–1.08)	0.91 (0.70–1.20)
P-trend				0.17	0.51

[1] Adjusted for age and sex (for all subjects' models). [2] Adjusted for age, baseline body mass index, pack-years of smoking, physical activity, prevalent diabetes, supplement use, and sex (for all subjects' models). MSDP, Mediterranean style dietary pattern; PY, person-years; and Ref, reference category.

Figures 1 and 2 show results of analyses exploring whether the risk estimates for total cancer were modified by anthropometric measures of body fat, cigarette smoking, or alcohol use in women and men. In each of these analyses, we explored the effects of a higher (≥19) vs. lower (<19) MSDP score combined with categories (some of which are sex-specific) of a second risk factor: For example, for BMI, we classified subjects into one of the following four categories: (1) low MSDP and high BMI (ref group), (2) low MSDP and low BMI, (3) high MSDP and high BMI, and (4) high MSDP and low BMI. Those in the fourth category were hypothesized to have the lowest total cancer risk. Among women, the HRs for categories 2, 3, and 4, respectively, were 1.14 (95% CI: 0.74–1.74), 0.84 (95% CI: 0.59–1.21), and 0.69 (95% CI: 0.47–1.01), suggesting that those with a higher MSDP score and a lower BMI did in fact have the lowest risk of cancer and that protective effect appeared more than additive. Similarly, women with higher MSDP scores who also had a lower WHtR had lower cancer risks than those with either risk factor alone. However, the excess risks due to the interaction of these adiposity factors and MSDP adherence were not statistically significant (p-interaction > 0.05).

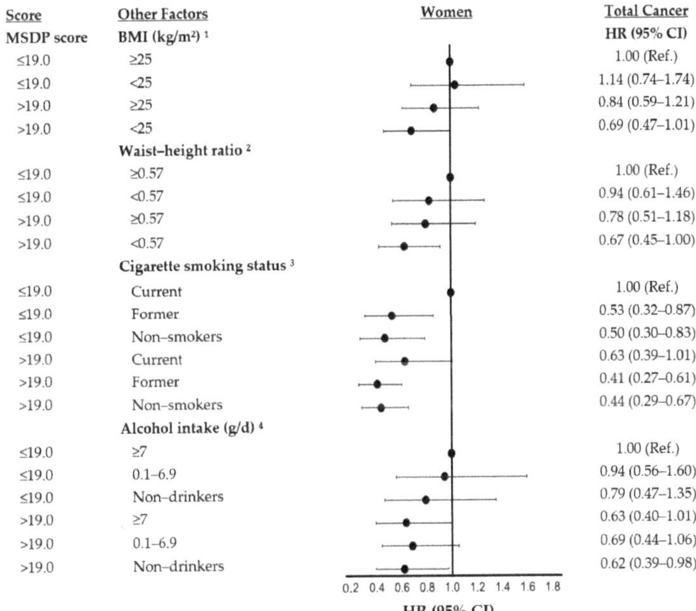

Figure 1. Independent and combined effects of MSDP adherence and anthropometric, lifestyle, and dietary factors on total cancer risk in women of the Framingham Offspring Study. [1] Models assessing effect modification by BMI were adjusted for age, pack-years of smoking, physical activity, prevalent diabetes, and supplement use. [2] Models assessing effect modification by waist-to-height ratio were adjusted for age, hip girth, pack-years of smoking, physical activity, prevalent diabetes, and supplement use. [3] Models assessing effect modification by smoking status were adjusted for age, BMI, physical activity, prevalent diabetes, and supplement use. [4] Models assessing effect modification by alcohol intake were adjusted for age, BMI, pack-years, physical activity, prevalent diabetes, and supplement use. For these models, wine was excluded from the MSDP score. MSDP, Mediterranean style dietary pattern; Ref, reference category; BMI, body mass index; and D, day.

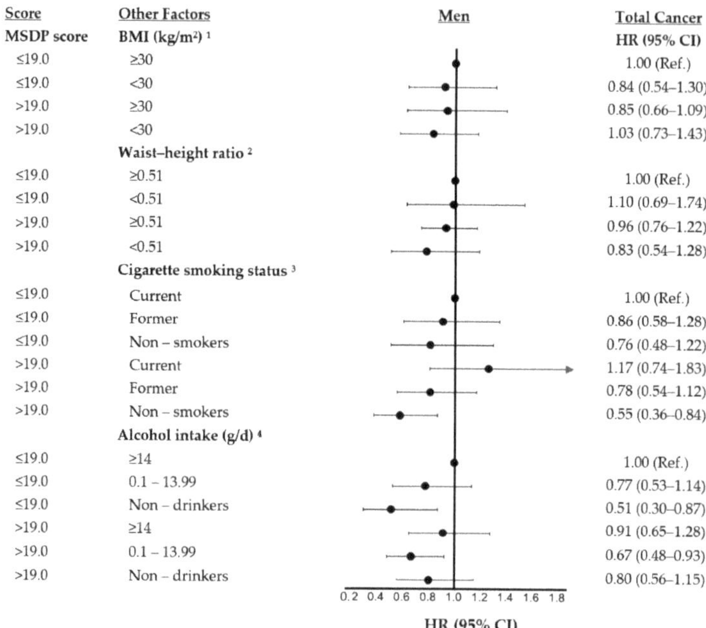

Figure 2. Independent and combined effects of MSDP adherence and anthropometric, lifestyle, and dietary factors on total cancer risk in men of the Framingham Offspring Study. [1] Models assessing effect modification by BMI were adjusted for age, pack-years of smoking, physical activity, prevalent diabetes, and supplement use. [2] Models assessing effect modification by waist-to-height ratio were adjusted for age, hip girth, pack-years of smoking, physical activity, prevalent diabetes, and supplement use. [3] Models assessing effect modification by smoking status were adjusted for age, BMI, physical activity, prevalent diabetes, and supplement use. [4] Models assessing effect modification by alcohol intake were adjusted for age, BMI, pack-years, physical activity, prevalent diabetes, and supplement use. For these models, wine was excluded from the MSDP score. Participants (n = 4, men) exceeding 100 g of alcohol per day were excluded from this model. MSDP, Mediterranean style dietary pattern; Ref, reference category; BMI, body mass index; and D, day.

We also explored possible effect modification on an additive scale of the MSDP scores by cigarette smoking and alcohol consumption. Non-smoking and former-smoking women generally had lower cancer risks, and these associations were slightly stronger among those with higher MSDP scores. In men, we found that those with higher MSDP scores who were non-smokers had the lowest risk of cancer (HR: 0.55; 95% CI: 0.36–0.84). Once again, while these effects were more than additive, the estimated interaction did not reach statistical significance (p-interaction > 0.05). Finally, there was no consistent evidence of effect modification by alcohol intake in either men or women.

Table 4 shows the sex-specific associations between categories of intake in each MSDP food group and total cancer risk. Compared with lower intakes, moderate intakes of some foods were associated with lower cancer risks, but these risks tended not to be statistically significant. For example, moderate dairy consumption was associated with a reduced risk of cancer in men (HR: 0.69, 95% CI: 0.51–0.94) and women (HR: 0.67, 95% CI: 0.47–0.96), but the p-values for trend were not statistically significant. Additionally, women with moderate (vs. lower) intakes of fruit, eggs, and potatoes and those who used olive oil tended to have lower cancer risks, and men with moderate intakes of potatoes tended to have lower risks.

Table 4. Hazard ratios for total cancer, according to the intake of each food group considered in the MSDP score in women and men of the Framingham Offspring Study.

Food Groups Considered in the MSDP Score (servings/day)	Women	Men
	HR (95% CI) [1]	HR (95% CI) [1]
Whole grains		
Low (<0.5)	1.00 (Ref.)	1.00 (Ref.)
Moderate (0.5–<1.0)	0.90 (0.62–1.31)	1.21 (0.88–1.66)
High (≥1)	1.11 (0.79–1.55)	1.10 (0.83–1.46)
P-trend	0.39	0.66
Fruit		
Low (<0.75)	1.00 (Ref.)	1.00 (Ref.)
Moderate (0.75–<2.5)	0.62 (0.44–0.88)	1.10 (0.81–1.48)
High (≥2.5)	0.87 (0.56–1.34)	1.02 (0.70–1.48)
P-trend	0.92	0.93
Vegetables		
Low (<1.5)	1.00 (Ref.)	1.00 (Ref.)
Moderate (1.5–<3)	0.99 (0.67–1.44)	0.95 (0.71–1.27)
High (≥3)	0.81 (0.52–1.26)	1.08 (0.75–1.55)
P-trend	0.22	0.59
Dairy		
Low (<0.5)	1.00 (Ref.)	1.00 (Ref.)
Moderate (0.5–<1.5)	0.67 (0.47–0.96)	0.69 (0.51–0.94)
High (≥1.5)	0.73 (0.49–1.08)	1.02 (0.74–1.40)
P-trend	0.43	0.17
Wine		
Non–drinkers	1.00 (Ref.)	1.00 (Ref.)
Drinkers	1.07 (0.81–1.41)	1.12 (0.89–1.42)
P-trend	0.64	0.34
Servings/week		
Olives, pulses & nuts		
Low (<0.5)	1.00 (Ref.)	1.00 (Ref.)
Moderate (0.5–<1.5)	1.09 (0.74–1.63)	0.80 (0.54–1.17)
High (≥1.5)	0.89 (0.58–1.37)	0.89 (0.59–1.34)
P-trend	0.32	0.82
Potatoes		
Low (<1.5)	1.00 (Ref.)	1.00 (Ref.)
Moderate (1.5–<3.5)	0.73 (0.50–1.08)	0.72 (0.51–1.01)
High (≥3.5)	0.76 (0.52–1.12)	0.87 (0.63–1.20)
P-trend	0.39	0.95
Poultry		
Low (<2)	1.00 (Ref.)	1.00 (Ref.)
Moderate (2–<5)	1.04 (0.66–1.64)	0.76 (0.54–1.08)
High (≥5)	1.19 (0.86–1.65)	0.88 (0.67–1.15)
P-trend	0.28	0.45
Meat		
Low (<2)	1.00 (Ref.)	1.00 (Ref.)
Moderate (2–<5)	1.22 (0.85–1.75)	1.63 (1.11–2.42)
High (≥5)	1.14 (0.75–1.74)	1.46 (0.95–2.26)
P-trend	0.81	0.67
Sweets		
Low (<7)	1.00 (Ref.)	1.00 (Ref.)
Moderate (7–<21)	0.92 (0.67–1.27)	0.96 (0.70–1.34)
High (≥21)	0.79 (0.50–1.27)	0.94 (0.64–1.37)
P-trend	0.34	0.75

Table 4. *Cont.*

Food Groups Considered in the MSDP Score (servings/day)	Women	Men
	HR (95% CI) [1]	HR (95% CI) [1]
Fish & other seafood		
Low (<2)	1.00 (Ref.)	1.00 (Ref.)
Moderate (2–<4)	0.84 (0.61–1.16)	1.12 (0.86–1.45)
High (≥4)	1.18 (0.83–1.67)	0.95 (0.69–1.32)
P-trend	0.41	0.85
Eggs		
Low (0)	1.00 (Ref.)	1.00 (Ref.)
Moderate (0.5–<1.5)	0.69 (0.48–0.98)	0.99 (0.70–1.40)
High (≥1.5)	0.82 (0.58–1.15)	0.97 (0.70–1.34)
P-trend	0.89	0.85
Olive oil		
No use of olive oil	1.00 (Ref.)	1.00 (Ref.)
Olive oil use	0.71 (0.49–1.01)	0.89 (0.67–1.20)
Olive oil and Vegetable oil use	0.93 (0.47–1.84)	1.37 (0.69–2.70)
P-trend	0.06	0.51

[1] Adjusted for age, body mass index, pack-years of smoking, physical activity, diabetes status, supplement use status, calorie intake and mutually adjusted for all the other MSDP food groups. MSDP, Mediterranean style dietary pattern and Ref., reference category.

4. Discussion

This is the first long-term population-based study using the MSDP score developed by Rumawas et al. [17] to assess the association between adherence to a Mediterranean-style diet and overall cancer risk. We observed that consuming a diet consistent with the principles of the Mediterranean diet was associated with reductions in total cancer risk among healthy adults aged 30 years or older but especially among women. Those women with moderate and higher MSDP scores had at least a 26% lower cancer risk. Further, non-overweight women with a higher MSDP score had a much lower risk of cancer than any other group of women. Among men, the effects of a Mediterranean-style diet were modified by smoking status in that non-smoking men with higher adherence to a Mediterranean-style diet had the lowest cancer risks. This study shows that the effects of a Mediterranean-style diet on cancer risk appear to be modified by other risk factors.

Only two previous studies have reported beneficial effects of the Mediterranean diet on total cancer incidence [14,15]. The multicenter EPIC study followed 142,605 men and 335,873 women for a median of 8.7 years and found that the highest Mediterranean diet score category was associated with 7% reductions in total cancer risk in both men and women [15]. These risk reductions were similar among participants in Mediterranean and non-Mediterranean countries, suggesting that the cancer-protective effects of this dietary pattern are not unique to countries in the Mediterranean region. We had insufficient power to detect statistically significant risk reductions of that magnitude in men in this current study, while the effect estimates (9% lower cancer risk associated with higher MSDP scores) were similar.

Our results suggested a stronger risk reduction in Framingham women than was observed in the larger EPIC study [15]. It is possible that the different approaches to scoring adherence to the Mediterranean diet might explain some of these differences in the observed effects. Specifically, the EPIC study used a diet score [16] with cut-off values for adequate intakes in European countries based on sex-specific population median food and nutrient intakes in those countries rather than intakes recommended by the Mediterranean diet pyramid (as was done in the current study). It may be that the lower baseline risk in the EPIC study led to weaker risk reductions in those countries. Other differences between the two studies included the addition of total alcohol in the score rather than red wine alone, as well as between-country differences in the food sources of nutrients such as monounsaturated fatty acids (MUFAs).

A separate analysis of data from the Greek cohort in the EPIC study found that adherence to a Mediterranean diet was associated with a statistically significant 22% lower risk of cancer overall, with stronger effects found among women (a statistically significant 27% lower risk) than among men, who had a non-statistically significant 17% lower risk [14]. It is possible that these stronger effects in the Greek EPIC cohort may have been due to the median-based scoring system better capturing Mediterranean diet adherence in that population than it did in more northern European countries. Our findings using the MSDP score were similar to the sex-specific differences found in the Greek EPIC cohort.

Excess body fat, including overweight, obesity, and weight gain (during middle adult years), are strong modifiable risk factors for certain cancers such as breast (post-menopausal), colorectal, and reproductive cancers [5,29]. The most common cancers among FOS participants were obesity-related cancers, mainly breast (post-menopausal) and colorectal. Prior analyses in the FOS study showed that gaining $\geq$0.45 kg ($\geq$1.0 pound) per year over ~14 years of follow-up during the middle-adult years increased the risk of overweight and obesity-related cancers by 38% (95% CI: 1.09–1.76) [23]. Data from the Nurses' Health Study and the Health Professional Follow-Up Study [30] as well as the Women's Health Initiative [31] also demonstrated that weight gain increased the risk of obesity-related cancers in women. Few studies have examined effect modification of the relation between a Mediterranean diet and cancer risk by baseline BMI [32,33], and results have been inconsistent. The current study in which the effects of a Mediterranean-style diet were stronger among leaner women differ from those of some studies in which the protective effects of the Mediterranean diet (e.g., on breast cancer risk) were found mainly among obese women [34,35].

In this study, the beneficial effect of a higher MSDP score in men was strongest among never-smokers, while in an earlier EPIC study, the beneficial effects of a Mediterranean-style diet were slightly stronger among current smokers than past or never smokers [15]. It is possible that these conflicting results may be due to differences in the baseline prevalence of cigarette smoking as well as rates of tobacco-related cancers between studies or that patterns of smoking may be differently associated with other behavioral risk factors in different studies.

An analysis of each food group considered in a Mediterranean diet score in association with total cancer risk has only been done once before. Our results showed that there was some tendency for certain food groups to be associated with lower cancer risks, especially in women, but none had statistically significant linear trends. In the earlier EPIC study, there was no evidence of an association between any individual component of the Mediterranean diet score with risk of total cancer [15]. The cancer-protective effect of the overall Mediterranean diet pattern is likely to be stronger than that of individual foods and nutrients associated with the diet pattern. The unique matrix of foods and nutrients as part of a Mediterranean dietary pattern may act synergistically to protect against the occurrence or spread of certain cancers [15]. Overall, the Mediterranean diet is rich in phytochemicals including polyphenols (such as flavonoids and resveratrol), carotenoids, and fiber, and its fatty acid profile is high in omega-3 and MUFAs. This overall nutrient profile is thought to have antioxidant, anti-inflammatory, antiproliferative, and antimutagenic properties [4–7,13,36–38].

In addition, the beneficial effects of the Mediterranean diet could be explained by intermediate effects on body fat or body composition resulting from the obesity-protective effect of the Mediterranean diet [5,10,39]. Despite the higher fat content of the Mediterranean diet, clinical and epidemiological studies showed that this diet has been linked with a low-to-moderate weight loss and lower abdominal adiposity [40–42]. A previous prospective analysis of individuals without diabetes mellitus in the FOS found that higher (vs. lower) MSDP scores were associated with lower waist circumference, less insulin resistance, and lower fasting plasma glucose and triglycerides, as well as higher HDL cholesterol [42]. One clinical trial demonstrated that people with metabolic syndrome assigned to a Mediterranean diet group had a 2.0 cm greater decrease in waist circumference and 2.8 kg greater weight loss than those in the control group (prudent diet) [40]. Further, central adiposity and metabolic dysfunction are both risk factors for cancer development. Therefore, a Mediterranean-style diet may result in less adiposity and metabolic dysfunction, thereby explaining in part the cancer-protective effects of this diet pattern.

The strengths of this study include its prospective design with carefully adjudicated cancer outcomes using standardized procedures [20]. In addition, most potential confounding or effect modifying variables were measured rather than self-reported in this study. In terms of limitations, because the score that we used assigned equal weight to each component, it assumes that the biological effects of these components are all equal with respect to different types of cancer, and this may not be the case. Another limitation is the limited distribution and relatively low average MSDP scores in this study. The highest score observed was 50.95, and the mean score was less than half that amount. Further, the FFQ used to calculate the MSDP score has a limited ability to estimate energy intake, thus making it a possible source of error. However, earlier validation studies of the FFQ used in these analyses found that many of the foods included in the MSDP score were adequately measured when compared with intakes from diet records [43]. A further limitation of this study is its sample size. The relatively small numbers of cancer cases gave us limited power to evaluate individual cancers, particularly for men and women, separately. This was especially problematic in the assessment of effect modification. Similarly, the food group analyses should be interpreted with caution due to the limited number of cases in some categories of intake and the strong correlations between different food groups in these analyses. Finally, the FOS subjects were exclusively Caucasian, so these results may not be representative of risks in a multiethnic population.

5. Conclusions

This is the first prospective cohort study to show that consumption of a Mediterranean-style diet may be one effective strategy for reducing total cancer risk in the US. In this study, the cancer-protective effects of higher Mediterranean diet adherence were strongest in women with less adiposity and among men who did not smoke. Given the high rates of cancer in the US [44,45], these findings have the potential to benefit large numbers of people.

Supplementary Materials: The following are available online at https://www.mdpi.com/article/10.3390/nu13114064/s1, Figure S1: Flowchart of study participants.

Author Contributions: Conceptualization, I.Y. and L.L.M.; methodology, I.Y., L.L.M., and M.R.S.; software, I.Y. and M.R.S.; validation, I.Y. and L.L.M.; formal analysis, I.Y.; investigation, I.Y., M.R.S.; P.F.J., V.X., R.C.E., and L.L.M.; resources, L.L.M.; data curation, M.R.S.; writing—original draft preparation, I.Y. and L.L.M.; writing—Review and Editing, I.Y., M.R.S., P.F.J., V.X., R.C.E., and L.L.M.; visualization, I.Y.; supervision, L.L.M.; project administration, I.Y.; funding acquisition, L.L.M. All authors have read and agreed to the published version of the manuscript.

Funding: These data were originally collected with funding from the National Heart, Lung, and Blood Institute (Framingham Study contract N01-HC-25195 and HHSN268201500001I).

Institutional Review Board Statement: The Framingham Offspring Study was conducted according to the guidelines of the Declaration of Helsinki and approved by the Institutional Review Board of Boston University Medical Center (protocol code H-32132, 1/5/2021). These analyses were approved by Institutional Review Board of the Boston University Medical Center (protocol code H-32086, original approval 01/30/2014).

Informed Consent Statement: Informed consent was obtained from all subjects involved in the study.

Data Availability Statement: This manuscript was prepared using Framingham Offspring Research Materials obtained from the NHLBI Biologic Specimen and Data Repository Information Coordinating Center and does not necessarily reflect the opinions or views of the Framingham Offspring Study or the NHLBI. Data can be found upon request at https://biolincc.nhlbi.nih.gov/studies/framoffspring/ (accessed on 12 November 2021).

Conflicts of Interest: The authors have declared that no conflicts of interest exist.

References

1. World Cancer Research Fund/American Institute for Cancer Research. Continuous Update Project Expert Report 2018. Diet, Nutrition, Physical Activity and Colorectal Cancer. Available online: Dietandcancerreport.Org (accessed on 1 October 2019).
2. Romagnolo, D.F.; Selmin, O.I. Mediterranean Diet and Prevention of Chronic Diseases. *Nutr. Today* **2017**, *52*, 208–222. [CrossRef] [PubMed]
3. Grosso, G.; Buscemi, S.; Galvano, F.; Mistretta, A.; Marventano, S.; Vela, V.; Drago, F.; Gangi, S.; Basile, F.; Biondi, A. Mediterranean Diet and Cancer: Epidemiological Evidence and Mechanism of Selected Aspects. *BMC Surg.* **2013**, *13*, S14. [CrossRef] [PubMed]
4. Kontou, N.; Psaltopoulou, T.; Panagiotakos, D.; Dimopoulos, M.A.; Linos, A. The Mediterranean Diet in Cancer Prevention: A Review. *J. Med. Food* **2011**, *14*, 1065–1078. [CrossRef] [PubMed]
5. Kwan, H.Y.; Chao, X.; Su, T.; Fu, X.; Tse, A.K.W.; Fong, W.F.; Yu, Z.-L. The Anticancer and Antiobesity Effects of Mediterranean Diet. *Crit. Rev. Food Sci. Nutr.* **2017**, *57*, 82–94. [CrossRef] [PubMed]
6. D'Alessandro, A.; De Pergola, G.; Silvestris, F. Mediterranean Diet and Cancer Risk: An Open Issue. *Int. J. Food Sci. Nutr.* **2016**, *67*, 593–605. [CrossRef]
7. Di Daniele, N.; Noce, A.; Vidiri, M.F.; Moriconi, E.; Marrone, G.; Annicchiarico-Petruzzelli, M.; D'Urso, G.; Tesauro, M.; Rovella, V.; De Lorenzo, A. Impact of Mediterranean Diet on Metabolic Syndrome, Cancer and Longevity. *Oncotarget* **2017**, *8*, 8947–8979. [CrossRef]
8. Rosato, V.; Guercio, V.; Bosetti, C.; Negri, E.; Serraino, D.; Giacosa, A.; Montella, M.; La Vecchia, C.; Tavani, A. Mediterranean Diet and Colorectal Cancer Risk: A Pooled Analysis of Three Italian Case–Control Studies. *Br. J. Cancer* **2016**, *115*, 862–865. [CrossRef]
9. van den Brandt, P.A.; Schulpen, M. Mediterranean Diet Adherence and Risk of Post-menopausal Breast Cancer: Results of a Cohort Study and Meta-Analysis. *Int. J. Cancer* **2017**, *140*, 2220–2231. [CrossRef]
10. Buckland, G.; Travier, N.; Cottet, V.; González, C.A.; Luján-Barroso, L.; Agudo, A.; Trichopoulou, A.; Lagiou, P.; Trichopoulos, D.; Peeters, P.H.; et al. Adherence to the Mediterranean Diet and Risk of Breast Cancer in the European Prospective Investigation into Cancer and Nutrition Cohort Study. *Int. J. Cancer* **2013**, *132*, 2918–2927. [CrossRef]
11. Jones, P.; Cade, J.E.; Evans, C.E.; Hancock, N.; Greenwood, D.C. The Mediterranean Diet and Risk of Colorectal Cancer in the UK Women's Cohort Study. *Int. J. Epidemiol.* **2017**, *46*, 1786–1796. [CrossRef]
12. Agnoli, C.; Grioni, S.; Sieri, S.; Palli, D.; Masala, G.; Sacerdote, C.; Vineis, P.; Tumino, R.; Giurdanella, M.C.; Pala, V.; et al. Italian Mediterranean Index and Risk of Colorectal Cancer in the Italian Section of the EPIC Cohort. *Int. J. Cancer* **2013**, *132*, 1404–1411. [CrossRef] [PubMed]
13. Ostan, R.; Lanzarini, C.; Pini, E.; Scurti, M.; Vianello, D.; Bertarelli, C.; Fabbri, C.; Izzi, M.; Palmas, G.; Biondi, F.; et al. Inflammaging and Cancer: A Challenge for the Mediterranean Diet. *Nutrients* **2015**, *7*, 2589–2621. [CrossRef] [PubMed]
14. Benetou, V.; Trichopoulou, A.; Orfanos, P.; Naska, A.; Lagiou, P.; Boffetta, P.; Trichopoulos, D. Conformity to Traditional Mediterranean Diet and Cancer Incidence: The Greek EPIC Cohort. *Br. J. Cancer* **2008**, *99*, 191–195. [CrossRef]
15. Couto, E.; Boffetta, P.; Lagiou, P.; Ferrari, P.; Buckland, G.; Overvad, K.; Dahm, C.C.; Tjønneland, A.; Olsen, A.; Clavel-Chapelon, F.; et al. Mediterranean Dietary Pattern and Cancer Risk in the EPIC Cohort. *Br. J. Cancer* **2011**, *104*, 1493–1499. [CrossRef] [PubMed]
16. Trichopoulou, A.; Orfanos, P.; Norat, T.; Bueno-de-Mesquita, B.; Ocké, M.C.; Peeters, P.H.; van der Schouw, Y.T.; Boeing, H.; Hoffmann, K.; Boffetta, P.; et al. Modified Mediterranean Diet and Survival: EPIC-Elderly Prospective Cohort Study. *BMJ* **2005**, *330*, 991. [CrossRef]
17. Rumawas, M.E.; Dwyer, J.T.; Mckeown, N.M.; Meigs, J.B.; Rogers, G.; Jacques, P.F. The Development of the Mediterranean-Style Dietary Pattern Score and Its Application to the American Diet in the Framingham Offspring Cohort. *J. Nutr.* **2009**, *139*, 1150–1156. [CrossRef]
18. Rimm, E.B.; Giovannucci, E.L.; Stampfer, M.J.; Colditz, G.A.; Litin, L.B.; Willett, W.C. Reproducibility and Validity of an Expanded Self-Administered Semiquantitative Food Frequency Questionnaire among Male Health Professionals. *Am. J. Epidemiol.* **1992**, *135*, 1114–1126. [CrossRef]
19. Ministry of Health and Welfare Supreme Scientific Health Council of Greece. Dietary Guidelines for Adults in Greece. *Arch. Hell. Med.* **1999**, *16*, 516–524.
20. Kreger, B.E.; Splansky, G.L.; Schatzkin, A. The Cancer Experience in the Framingham Heart Study Cohort. *Cancer* **1991**, *67*, 1–6. [CrossRef]
21. Moore, L.L.; Bradlee, M.L.; Singer, M.R.; Splansky, G.L.; Proctor, M.H.; Ellison, R.C.; Kreger, B.E. BMI and Waist Circumference as Predictors of Lifetime Colon Cancer Risk in Framingham Study Adults. *Int. J. Obes.* **2004**, *28*, 559–567. [CrossRef]
22. Moore, L.L.; Chadid, S.; Singer, M.R.; Kreger, B.E.; Denis, G.V. Metabolic Health Reduces Risk of Obesity-Related Cancer in Framingham Study Adults. *Cancer Epidemiol. Biomark. Prev.* **2014**, *23*, 2057–2065. [CrossRef] [PubMed]
23. Chadid, S.; Singer, M.R.; Kreger, B.E.; Bradlee, M.L.; Moore, L.L. Midlife Weight Gain Is a Risk Factor for Obesity-Related Cancer. *Br. J. Cancer* **2018**, *118*, 1665–1671. [CrossRef]
24. Mustafa, J.; Ellison, R.C.; Singer, M.R.; Bradlee, M.L.; Kalesan, B.; Holick, M.F.; Moore, L.L. Dietary Protein and Preservation of Physical Functioning Among Middle-Aged and Older Adults in the Framingham Offspring Study. *Am. J. Epidemiol.* **2018**, *187*, 1411–1419. [CrossRef] [PubMed]

25. Kannel, W.B.; Belanger, A.; D'Agostino, R.; Israel, I. Physical Activity and Physical Demand on the Job and Risk of Cardiovascular Disease and Death: The Framingham Study. *Am. Heart J.* **1986**, *112*, 820–825. [CrossRef]
26. Rothman, K.J.; Greenland, S.; Walker, A.M. Concepts of Interaction. *Am. J. Epidemiol.* **1980**, *112*, 467–470. [CrossRef] [PubMed]
27. Knol, M.J.; VanderWeele, T.J.; Groenwold, R.H.H.; Klungel, O.H.; Rovers, M.M.; Grobbee, D.E. Estimating Measures of Interaction on an Additive Scale for Preventive Exposures. *Eur. J. Epidemiol.* **2011**, *26*, 433–438. [CrossRef]
28. Li, R.; Chambless, L. Test for Additive Interaction in Proportional Hazards Models. *Ann. Epidemiol.* **2007**, *17*, 227–236. [CrossRef]
29. Giacosa, A.; Barale, R.; Bavaresco, L.; Gatenby, P.; Gerbi, V.; Janssens, J.; Johnston, B.; Kas, K.; La Vecchia, C.; Mainguet, P.; et al. Cancer Prevention in Europe: The Mediterranean Diet as a Protective Choice. *Eur. J. Cancer Prev.* **2013**, *22*, 90. [CrossRef]
30. Zheng, Y.; Manson, J.E.; Yuan, C.; Liang, M.H.; Grodstein, F.; Stampfer, M.J.; Willett, W.C.; Hu, F.B. Associations of Weight Gain from Early to Middle Adulthood With Major Health Outcomes Later in Life. *JAMA* **2017**, *318*, 255–269. [CrossRef]
31. Arnold, M.; Jiang, L.; Stefanick, M.L.; Johnson, K.C.; Lane, D.S.; LeBlanc, E.S.; Prentice, R.; Rohan, T.E.; Snively, B.M.; Vitolins, M.; et al. Duration of Adulthood Overweight, Obesity, and Cancer Risk in the Women's Health Initiative: A Longitudinal Study from the United States. *PLOS Med.* **2016**, *13*, e1002081. [CrossRef]
32. Bosetti, C.; Turati, F.; Pont, A.D.; Ferraroni, M.; Polesel, J.; Negri, E.; Serraino, D.; Talamini, R.; Vecchia, C.L.; Zeegers, M.P. The Role of Mediterranean Diet on the Risk of Pancreatic Cancer. *Br. J. Cancer* **2013**, *109*, 1360–1366. [CrossRef] [PubMed]
33. Filomeno, M.; Bosetti, C.; Bidoli, E.; Levi, F.; Serraino, D.; Montella, M.; La Vecchia, C.; Tavani, A. Mediterranean Diet and Risk of Endometrial Cancer: A Pooled Analysis of Three Italian Case-Control Studies. *Br. J. Cancer* **2015**, *112*, 1816–1821. [CrossRef] [PubMed]
34. Murtaugh, M.A.; Sweeney, C.; Giuliano, A.R.; Herrick, J.S.; Hines, L.; Byers, T.; Baumgartner, K.B.; Slattery, M.L. Diet Patterns and Breast Cancer Risk in Hispanic and Non-Hispanic White Women: The Four-Corners Breast Cancer Study. *Am. J. Clin. Nutr.* **2008**, *87*, 978–984. [CrossRef] [PubMed]
35. Turati, F.; Carioli, G.; Bravi, F.; Ferraroni, M.; Serraino, D.; Montella, M.; Giacosa, A.; Toffolutti, F.; Negri, E.; Levi, F.; et al. Mediterranean Diet and Breast Cancer Risk. *Nutrients* **2018**, *10*, 326. [CrossRef]
36. Demetriou, C.A.; Hadjisavvas, A.; Loizidou, M.A.; Loucaides, G.; Neophytou, I.; Sieri, S.; Kakouri, E.; Middleton, N.; Vineis, P.; Kyriacou, K. The Mediterranean Dietary Pattern and Breast Cancer Risk in Greek-Cypriot Women: A Case-Control Study. *BMC Cancer* **2012**, *12*, 113. [CrossRef]
37. Schwingshackl, L.; Hoffmann, G. Does a Mediterranean-Type Diet Reduce Cancer Risk? *Curr. Nutr. Rep.* **2016**, *5*, 9–17. [CrossRef]
38. Toledo, E.; Salas-Salvadó, J.; Donat-Vargas, C.; Buil-Cosiales, P.; Estruch, R.; Ros, E.; Corella, D.; Fitó, M.; Hu, F.B.; Arós, F.; et al. Mediterranean Diet and Invasive Breast Cancer Risk Among Women at High Cardiovascular Risk in the PREDIMED Trial: A Randomized Clinical Trial. *JAMA Intern. Med.* **2015**, *175*, 1752–1760. [CrossRef]
39. Bendall, C.L.; Mayr, H.L.; Opie, R.S.; Bes-Rastrollo, M.; Itsiopoulos, C.; Thomas, C.J. Central Obesity and the Mediterranean Diet: A Systematic Review of Intervention Trials. *Crit. Rev. Food Sci. Nutr.* **2018**, *58*, 3070–3084. [CrossRef]
40. Esposito, K.; Marfella, R.; Ciotola, M.; Di Palo, C.; Giugliano, F.; Giugliano, G.; D'Armiento, M.; D'Andrea, F.; Giugliano, D. Effect of a Mediterranean-Style Diet on Endothelial Dysfunction and Markers of Vascular Inflammation in the Metabolic Syndrome: A Randomized Trial. *JAMA* **2004**, *292*, 1440–1446. [CrossRef]
41. Nordmann, A.J.; Suter-Zimmermann, K.; Bucher, H.C.; Shai, I.; Tuttle, K.R.; Estruch, R.; Briel, M. Meta-Analysis Comparing Mediterranean to Low-Fat Diets for Modification of Cardiovascular Risk Factors. *Am. J. Med.* **2011**, *124*, 841–851.e2. [CrossRef]
42. Rumawas, M.E.; Meigs, J.B.; Dwyer, J.T.; McKeown, N.M.; Jacques, P.F. Mediterranean-Style Dietary Pattern, Reduced Risk of Metabolic Syndrome Traits, and Incidence in the Framingham Offspring Cohort123. *Am. J. Clin. Nutr.* **2009**, *90*, 1608–1614. [CrossRef] [PubMed]
43. Feskanich, D.; Rimm, E.B.; Giovannucci, E.L.; Colditz, G.A.; Stampfer, M.J.; Litin, L.B.; Willett, W.C. Reproducibility and Validity of Food Intake Measurements from a Semiquantitative Food Frequency Questionnaire. *J. Am. Diet Assoc.* **1993**, *93*, 790–796. [CrossRef]
44. Barak, Y.; Fridman, D. Impact of Mediterranean Diet on Cancer: Focused Literature Review. *Cancer Genom. Proteom.* **2017**, *14*, 403–408. [CrossRef]
45. Weir, H.K.; Thompson, T.D.; Soman, A.; Møller, B.; Leadbetter, S.; White, M.C. Meeting the Healthy People 2020 Objectives to Reduce Cancer Mortality, Peer Reviewed: Meeting the Healthy People 2020 Objectives to Reduce Cancer Mortality. *Prev. Chronic Dis.* **2015**, *12*, E104. [CrossRef] [PubMed]

MDPI

Review

Adherence to the Mediterranean Diet: Impact of Geographical Location of the Observations

Elisa Mattavelli [1], Elena Olmastroni [1], Daniela Bonofiglio [2,3], Alberico L. Catapano [1,4], Andrea Baragetti [1,4,†] and Paolo Magni [1,4,*,†]

1 Dipartimento di Scienze Farmacologiche e Biomolecolari, Università degli Studi di Milano, 20133 Milan, Italy; elisa.mattavelli@unimi.it (E.M.); elena.olmastroni@unimi.it (E.O.); alberico.catapano@unimi.it (A.L.C.); andrea.baragetti@unimi.it (A.B.)
2 Department of Pharmacy, Health and Nutritional Sciences, University of Calabria, 87036 Rende, Italy; daniela.bonofiglio@unical.it
3 Centro Sanitario, University of Calabria, 87036 Rende, Italy
4 IRCCS MultiMedica, Sesto San Giovanni, 20099 Milan, Italy
* Correspondence: paolo.magni@unimi.it; Tel.: +39-02-50318229
† These authors contributed equally to this work.

Abstract: The Mediterranean diet has emerged as a comprehensive lifestyle, including specific foods and meal composition and a set of behavioural and social features. Adherence to the Mediterranean diet has been shown to promote health and reduce the prevalence of chronic diseases. The actual implementation of the Mediterranean diet is affected by several sociocultural factors as well as geographical components. Indeed, the geographical location, such as a specific country or different areas in a country and specific latitude and climate, appears to be an important factor that may strongly affect the implementation of the Mediterranean diet or some of its principles as well as the adherence to it. Another dynamic component affecting personal nutritional choices, also regarding adherence to the Mediterranean diet and its principles, is the individual life-long trajectory of food preference and nutrition habits and awareness. In this review, we discuss the current evidence on the impact of geographical location on adherence to the Mediterranean diet.

Keywords: Mediterranean diet; adherence; awareness; health; geographical location; ethnic; nutrition guidelines; chronic disease

Citation: Mattavelli, E.; Olmastroni, E.; Bonofiglio, D.; Catapano, A.L.; Baragetti, A.; Magni, P. Adherence to the Mediterranean Diet: Impact of Geographical Location of the Observations. *Nutrients* **2022**, *14*, 2040. https://doi.org/10.3390/nu14102040

Academic Editor: Jose Lara

Received: 28 February 2022
Accepted: 11 May 2022
Published: 13 May 2022

1. Introduction

Among the different nutritional approaches aimed at promoting and maintaining good health as well as preventing chronic diseases, the Mediterranean diet has emerged over the decades as a comprehensive lifestyle, that, in addition to specific foods [1] and meal composition [2,3], also includes a set of behavioural and social features such as the production of specific foods as well as social exchange and communication [4–7]. The Mediterranean diet reflects the food patterns typical of southern regions of Greece, Italy, Spain and France in the early 1960s, where adult life expectancy was quite high, and the prevalence of diet-related chronic diseases was low [4–6]. Such food patterns, based on fresh or minimally processed or refined foods, were more specifically present in rural areas with easier availability of such dietary components, compared to urban areas [8]. The nutritional characteristics of the Mediterranean diet have inspired institutions and experts in several regions of the world and, more specifically, in individual countries, to implement healthier approaches to food consumption into their specific nutritional guidelines [9]. Some examples of this approach include the New Nordic Diet [10] and the US Dietary Guidelines for Americans, 2020–2025, 9th ed. [11]. In this context, it is important to emphasize that, in order to be most effective, nutritional guidelines need to be developed according to several specific factors, such as local cultural, ethnic and traditional features, and geographic and socioeconomic components, as well as to the availability of specific local food products [9]. Among these features, the geographical location (i.e., a specific country, different areas in a country, such as seaside or mountainside or plains far from both, and specific latitude and climate, including climate changes, as well as the related social features) appears to be a crucial factor that may affect to a large extent the actual applicability of the Mediterranean diet or some of its principles as well as the actual adherence to it [12–18]. An additional dynamic component affecting personal nutritional choices, also regarding adherence to the Mediterranean diet and its principles, is the individual life-long trajectory of food preference and nutrition habits and awareness, including feeding behavior aspects [19]. In the past, this aspect has been associated with local traditions and ethnic features and therefore remained rather stable over time within a specific population. However, over the last decades, due to the exponential diffusion of multiple communication media, the modifications of such traditional nutrition habits have been more frequent, leading to either a nutritional transition to unhealthy habits, especially in younger subjects [17,20–22] and in selected areas such as the Middle Eastern and North African (MENA) region [12], or a greater awareness of the health advantages of a good adherence to the Mediterranean diet, especially in middle/older age subjects, with improved nutrition patterns during their lifetime [23]. Thus, the epidemiological and experimental observations reporting adherence to the Mediterranean diet and its impact on health issues should also take into consideration another component: the time-related longitudinal changes within a specific population.

This narrative review discusses the current evidence on the potential impact of these two important factors, geographical location and time-related life-long change, on adherence to the Mediterranean diet. To this aim, the PubMed and Excerpta Medica Database (Embase) were searched from inception until April 2022. The search was also extended to the gray literature. Used search terms, with a combination of MeSH terms if applicable in each database, included: "Mediterranean diet", "geographical location", "region", "regional", "adherence", "Nordic diet", and "age".

2. Adherence to the Mediterranean Diet and Chronic Disease Prevention

Several epidemiological observations and intervention studies demonstrated that implementation of the Mediterranean diet, either as spontaneous adherence [24,25] or following specific intervention approaches (i.e., Lyon Diet Heart Study [26] and PREDIMED Study [27]), are associated with numerous health benefits. These include a reduced incidence of cardiovascular diseases [26–30], including peripheral artery disease [31], diabetes mellitus [32] and the metabolic syndrome [33] and type 2 diabetes mellitus [34,35]. Moreover, adherence to the Mediterranean diet has been found to show positive effects on cancer incidence [36], possibly thanks to the presence of natural compounds with protective effects [37]. Greater adherence to the Mediterranean diet has also been associated with a reduced incidence of neurodegenerative diseases (Parkinson's and Alzheimer's disease) [38,39] and, coherently, of cognitive dysfunction and physical impairment [40–42].

3. Adherence to the Mediterranean Diet and Geographic Location in Adults

As an extension of the above-reported observations, it is important to associate the term "Mediterranean diet" not only with the food quality and meal composition, but also with a particular way of cooking, eating and more. The Mediterranean diet indeed involves a series of skills, knowledge, symbols, rituals, and traditions related to crops, harvesting, animal husbandry, fishing, conservation, processing, and, more specifically, to the sharing and consumption of food [7]. Eating together is the basis of the cultural identity and continuity of communities throughout the Mediterranean area, representing an important moment of social exchange and communication, which results in the reinforcement of the relationships within family, groups, and community [43]. For these reasons, the Mediterranean diet was inscribed in 2013 in the Representative List of the Intangible Cultural Heritage of Humanity Countries: Greece, Italy, Spain, Morocco, Cyprus, Croatia, and Portugal [7]. Based on these considerations, it appears that the specific geographical location and all the related features strongly impact the implementation of the Mediterranean diet or some of its aspects, on the extent of awareness of its benefits and, ultimately, on the adherence of a population. Interestingly, the relevant change in the society features and the associated globalization that occurred in the last half-century and are still ongoing impacted and still impact the relationship between local habits and adherence to the Mediterranean diet, sometimes in opposite directions. A very recent systematic analysis of adherence to the Mediterranean diet among adults in a large set of Mediterranean countries reported that most available studies (in the 2010–2021 period) are from European Mediterranean countries, with fewer studies from Mediterranean countries in North African and Middle Eastern regions [18]. In general, low or moderate adherence was reported by the different studies, without major sex and age differences.

The analysis of adherence to the Mediterranean diet in cross-sectional studies may be useful to unveil possible differences between populations living in the same country or even district, but in different geographical and socio-cultural conditions, such as, for example, in rural areas and in urban contexts, two quite diverse geographical settings. Table 1 reports the main outcomes of the studies conducted in adults and discussed in this section.

Table 1. Adherence to the Mediterranean diet and geographic location in adults.

Study Name	Country	Year	Scoring System	Adherence to Mediterranean Diet	Reference
PLIC	Italy (urban)	2017	PREDIMED	higher than PLIC-Chiesa	[44,45]
PLIC-Chiesa	Italy (mountain)	2020	PREDIMED	lower than PLIC	[46]
DIMERICA	Spain (different regions)	2016	MD adherence score	lowest in Southestern Spain	[47]
ATTICA	Greece (urban)	2021	MedDietScore	higher in women, older people	[48]
PERSEAS	Greece (island)	2017	MedDietScore	moderate, despite being in a small island	[49]
	Cyprus	2021	MedDietScore	higher in males/rural areas	[50]
	Spain	2022	PREDIMED	no major changes during COVID-19 lockdown	[51]
	Italy	2020	PREDIMED	no major changes during COVID-19 lockdown	[52]
	Australia	1999	composite score	higher was associated with longer survival	[53]
	USA (different regions)	2020	MD adherence score	significant geospatial/population disparities	[54]
	USA (Stroke Belt/other regions)	2019	PREDIMED	lower in Stroke Belt group	[55]

One example of this is offered by the comparison between two epidemiological studies recently conducted in northern Italy and rather far from the Mediterranean Sea. One study involved the PLIC cohort, conducted in the urban area of Milan, Italy, on 2500 adult volunteers [44,45] and the other involved the PLIC Chiesa cohort, conducted in about 800 adult subjects in an isolated village in the Italian Alps, at 1000 m above sea level [46]. These two populations, although living just 150 km apart, represent two socio-cultural heterogenous groups: one living in an urban context and the other being a rather homogeneous community with greater isolation and specific socio-cultural identity, respectively. A preliminary analysis of the adherence to the Mediterranean diet, assessed by the PREDIMED 14-item tool [56] (Table 2), was conducted in subgroups from the two studies, according to a case-control design (age- and sex-matched). It was found that PLIC-Chiesa individuals were less adherent to the Mediterranean diet than the corresponding PLIC subjects (mean PREDIMED score 7.38 vs. 8.22, respectively; $p < 0.001$), consistent with the traditional mountain diet, mostly consisting of food of animal origin. The full study will unveil whether this finding is associated with different health outcomes.

Table 2. Mediterranean diet score systems used in the reported studies.

Scoring System	Range for Classification in Categories		
	Low	Moderate	High
PREDIMED (MEDAS) [56]	≦5	6–9	≧10
MD adherence score [47]	<5	5–7	>7
MedDietScore [57]	<25/55	26–35/55	>35/55
KIDMED [58]	≦3	4–7	8
Mediterranean diet scale (MDS) [59,60]	0–3	4–5	6–9
Mediterranean Diet Serving Score (MDSS) [61]	0–9.9	10–13.9	≧14 (24 max)

Mediterranean diet scores were assessed by direct administration by a trained dietician (PREDIMED) or by evaluating a food frequency questionnaire (all others). The related references are indicated.

A regional variability of adherence to the Mediterranean diet (evaluated by the MD adherence score, Table 2) has also been reported in Spain, where a recent study conducted in all regions showed that southeastern Spain had the lowest score for adherence to the Mediterranean diet specifically related to low consumption of fish and plant products. A lower adherence score to the Mediterranean diet was also strongly associated with the prevalence of hypertension [47]. Studies on adherence to the Mediterranean diet and health outcomes in urban as well as rural areas have also been conducted in Greece. The ATTICA epidemiological study estimated the level of adherence to the Mediterranean diet in a sample (2749 participants) population of the Athens metropolitan area [48], by using the MedDietScore [57] (Table 2). Higher adherence to the Mediterranean diet was observed in areas with a greater proportion of women and older people with a lower unemployment rate and immigrant population, as well as in locations with more green areas and a higher frequency of supermarkets and street markets. On the contrary, adherence to the traditional Mediterranean diet, evaluated by the MedDietScore, in the population of the Elafonisos island, a small Greek island, was found to be moderate and associated with low physical activity and high prevalence of obesity and traditional risk factors for cardiovascular diseases, suggesting the need for lifestyle improvement programs in rural isolated areas of the Mediterranean basin [49]. Another study, conducted in the Republic of Cyprus, showed that adherence to the Mediterranean diet (assessed by the MedDietScore) in a sample of the general population was greater in males and residents of rural regions compared with females and residents of urban regions [50]. Interestingly, adherence to the Mediterranean diet, once consolidated, appears to be rather robust, even under challenging conditions, such as the recent COVID-19 pandemic. In this regard, studies reported that no major changes in Mediterranean diet adherence occurred in Spanish university students [51] and in a sample of the Italian population [52] during the recent COVID-19 lockdown, a major lifestyle challenge.

Outside the Mediterranean basin, certain data and experiences related to the Mediterranean diet are interesting. Longevity is a relevant advantage associated with adherence to the Mediterranean diet for a substantial part of life [62,63]. To address this issue in Australia, which is a country geographically and culturally quite far away from the Mediterranean area, a prospective cohort study including both Anglo-Celts and Greek-Australians was conducted in Melbourne, with the goal to evaluate whether adherence to the Mediterranean diet affected the survival of elderly people in this developed non-Mediterranean country. The authors reported that a diet that adheres to the principles of the Mediterranean diet was associated with longer survival among Australians of both Greek and Anglo-Celtic origin [53].

In the US, the Mediterranean diet, due to its palatability and its consequent high acceptability, is considered a good opportunity for dietary improvement, for example by increasing consumption of fresh vegetables, fruit, grains, and olive oil since the early 20th century [6]. Interestingly, improved awareness of and, consequently, adherence to the Mediterranean diet style in the US has also been promoted by populations that immigrated to the United States from Greece, Italy and Spain. In agreement with previous versions, the 2020–2025 version of the Dietary Guidelines for Americans [11] currently includes the Healthy Mediterranean-Style Dietary Pattern, which is considered a variation of the Healthy US-Style Dietary Pattern, based on the types and proportions of foods typically consumed by Americans, although in nutrient-dense forms and appropriate amounts. A very recent study conducted a geospatial analysis of Mediterranean diet adherence (assessed by the MD adherence score) in the US [54], collecting data across the US regions and exploring the predictive factors of such adherence among US adults (over 20,000 participants). High adherence was observed in 46.5% of the sample. Higher adherence clusters were mainly located in the western and northeastern coastal areas of the USA, whereas lower Mediterranean diet adherence clusters were observed in south and east-north-central regions. Being older, black, not a current smoker, having a college degree or above, an annual household income ≥USD US 75K, exercising ≥4 times/week and watching TV/video <4 h/d were each associated with higher odds of high adherence. The authors concluded that across the US regions there is a significant geospatial and population disparity in adherence to the Mediterranean diet, possibly leading to the greater prevalence of chronic diseases. In the US, although recommended by the current Dietary Guidelines for Americans, as indicated above, the Mediterranean diet is still poorly associated with its benefits, particularly in the Stroke Belt, a large 11-state region in the southeast part of the country with an unusually high incidence of stroke and other forms of cardiovascular disease. A recent study examined Mediterranean diet adherence (assessed by the PREDIMED score) and perceived knowledge, benefits, and barriers in the US [55]. Convenience, sensory factors and health were greater barriers to the Mediterranean diet in the Stroke Belt group, but not in the other groups/regions. Participants with a bachelor's degree or higher showed greater adherence to the Mediterranean diet, whereas obese participants had a lower adherence.

When considering the geographical location and Mediterranean diet, it is worthwhile to mention an important European experience, represented by the New Nordic Diet, developed in some northern Europe countries (Denmark, Sweden and Finland), characterized by a markedly colder climate. Here, the plant-based nutrition present in the Mediterranean diet is translated into the consumption of healthy regional-specific foods, such as vegetables available in that area (pears, apples, berries, root and cruciferous vegetables, cabbages, rye bread and whole grain) as well as potatoes, a high intake of fish, low-fat dairy products, and vegetable fats, among other dietary lipid sources [10,64,65]. Moreover, it contains 35% less meat than the average Danish diet and appears to be effective in sustainability terms [66]. Rather little research has been conducted so far on the long-term health effects of adherence to the New Nordic Diet. A recent study found a non-significant inverse association with the overall incidence of myocardial infarction and of stroke in men [67], whereas no association was observed with the incidence of type 2 diabetes mellitus. These findings suggest the need for further studies to highlight which components of the Mediterranean diet provide the well-established cardioprotective effect. In this context, one may speculate about the effects of the different fats used for cooking: olive oil in the Mediterranean diet vs. rapeseed oil in the New Nordic diet [68,69].

Figure 1 reports the main geographical features reported for adults and is discussed in this section.

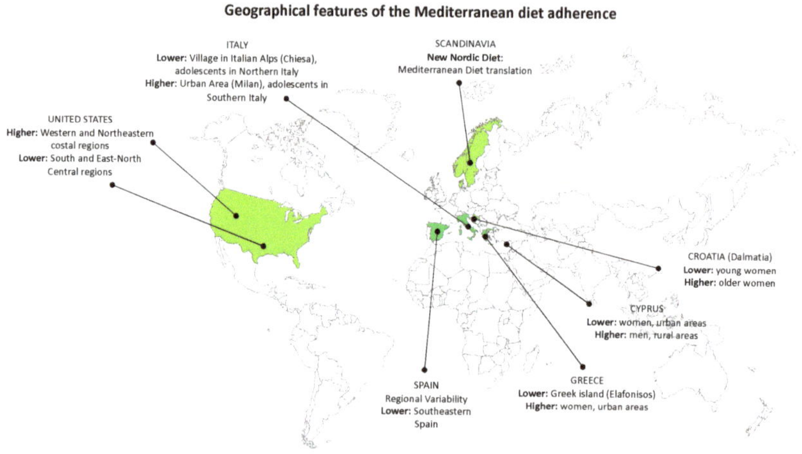

Figure 1. Geographical features of Mediterranean diet adherence in adults.

4. Adherence to the Mediterranean Diet and Geographic Location in Adolescents

Although several reports are available regarding adherence to the Mediterranean diet in the adult population, little information is available regarding children and adolescents, spontaneously adhering to this nutrition pattern early in life. Table 3 reports the main outcomes of the studies conducted in adolescents and discussed in this section.

Table 3. Adherence to the Mediterranean diet and geographic location in adolescents.

Study Name	Country	Year	Scoring System	Adherence to Mediterranean Diet	Reference
	Greece	2009	KIDMED	greater if good knowledge of Med diet	[70]
MDFS group	Italy	2014	Mediterr. diet scale	low: 38.6%; high: 14%. Higher in Southern Italy	[71]
DIMENU	Italy (Southern)	2020	KIDMED	medium: 60.9%	[72]
	Italy (Sicily)	2013	KIDMED	moderate in general, lower in urban settings	[73]
	Italy	2021	KIDMED	higher associates to antiinflammatory profile	[74]
DIMENU	Italy (Southern)	2021	KIDMED	higher with education programs	[75]
	Croatia (Dalmatia)	2021	MDSS	reduced in younger women	[61]

In the context of younger age, education and awareness about the Mediterranean diet and its principles are very important, as shown by a study conducted in Greece and showing that Greek adolescents consume a more westernized diet, which is quite detached from the traditional Mediterranean diet, although actual knowledge about the Mediterranean diet was associated with greater adherence to it [70], according to the evaluation by KIDMED scoring [58] (Table 2). Data from different Italian regions [71] (obtained using the Mediterranean diet scale (MDS) according to [59], revised by [60], Table 2) in a sample of 565 adolescents aged 12–19 years indicate that 38.6% of subjects had low adherence to the Mediterranean diet, whereas only 14% had high adherence. In this study, adolescents from the southern region of Italy showed the highest adherence, compared with those from the northeast and northwest Italy, which are mostly far from the Mediterranean Sea. Recently, in the DIMENU cross-sectional study, carried out in adolescents from the southern Italy area, a medium adherence to the Mediterranean diet assessed by the KIDMED score was reported (medium adherence 60.87%) with no significant differences according to gender [72]. Similar data were also reported by other authors among adolescents living in the Mediterranean region [73], and in the adult population from the same Mediterranean area in which a direct association with age was found [76]. Specifically, Caparello et al. reported that in adulthood the percentage of adherers to recommendations for fruit, nuts, and fish, estimated by the PREDIMED score, was below the dietary guidelines [76]. Supporting this concept, in adolescence, higher consumers of nuts, olive oil and fish showed a better serum metabolic profile, underlining the need to improve the consumption of distinctive components of the Mediterranean diet pattern and to encourage people to change their eating behaviour [77]. In addition, optimal Mediterranean diet adherence (assessed by KIDMED) in adolescents performing vigorous intensity levels of physical activity was found to be associated with lower lipid profile markers and reduced insulin concentrations, reinforcing the healthy benefits of the Mediterranean-style diet pattern [72]. Moreover, serum from adolescents who follow an optimal adherence to the Mediterranean diet (KIDMED) displayed anti-inflammatory and antioxidant properties which may provide chances for the prevention of metabolic and chronic diseases in adulthood [74]. However, a better knowledge of the composition of the healthy Mediterranean diet pattern, in terms of food sources of macro- and micronutrients through nutrition education programs is necessary to improve the adherence to the Mediterranean diet (KIDMED) in the young population. Indeed, the awareness towards the promotion of the Mediterranean diet as a global nutritionally balanced and healthy dietary pattern led to a decreased inflammatory status in adolescents of southern Italy [75].

An interesting "generation shift" from a Mediterranean diet pattern to a more "westernized" type of diet has also been observed in Dalmatia, the Mediterranean coastal part of Croatia, highlighting that younger women, although having a higher education and socioeconomic status, showed a lower adherence to the Mediterranean diet (assessed by the Mediterranean Diet Serving Score [61], Table 2) and a healthier lifestyle than older women from the same region [17].

5. Conclusions

This paper highlights the relevance of the geographical location and related social features in the actual adherence to the Mediterranean diet, thereby promoting (or not) several health advantages. Moreover, evidence is present that improved health and longevity results from prolonged exposure to this dietary pattern, emphasizing the time factor and the actual moment of its acquisition. However, in several instances, Mediterranean populations have been showing moderate adherence to the Mediterranean diet in the past 10 years [18], which suggests the need for improving adherence to the Mediterranean diet even in the countries of its origin. In the future, the ways in which a palatable and healthful dietary pattern, such as the Mediterranean diet, can be communicated and promoted need to be investigated beyond medical and nutrition authorities, for example, by employing culinary and marketing strategies. Importantly, several critical issues are present when adapting the Mediterranean diet to non-Mediterranean populations [78]. The approach to studying diet–health relationships has progressively moved from individual dietary components to overall dietary patterns that affect the interaction and balance of personal metabolome (low-molecular-weight metabolites) and microbiome (host-enteric microbial ecology). Future studies will be needed to address this complexity, unveiling how metabolome and microbiome profiles are modulated by high adherence to the Mediterranean diet in the different populations and regions to promote health and longevity [79,80].

Author Contributions: Conceptualization, P.M.; methodology, E.M., E.O.; software, E.O.; validation, P.M., D.B., A.L.C.; formal analysis, P.M., A.B.; investigation, P.M.; resources, P.M., A.L.C.; data curation, P.M., E.M.; writing—original draft preparation, P.M.; writing, review and editing, P.M., D.B., A.B., A.L.C.; project administration, P.M.; funding acquisition, P.M. All authors have read and agreed to the published version of the manuscript.

Funding: The work of A.L.C., A.B. and P.M. has been supported by Ministry of Health—Ricerca Corrente—IRCCS MultiMedica. The work of P.M. and A.L.C. is supported by the Ministry of Health-IRCCS MultiMedica GR-2016-02361198. The work of P.M. is also supported by the Universita' degli Studi di Milano (Transition grant PSR2015-1720PMAGN_01). The work of A.L.C. is supported by the Fondazione Cariplo 2015-0524 and 2015-0564; H2020 REPROGRAM PHC-03-2015/667837-2; ERANET ER-2017-2364981; PRIN 2017H5F943; Ministry of Health-IRCCS MultiMedica GR-2011-02346974; SISA Lombardia and Fondazione SISA. The work of A.B. is supported by a research grant from the Peanut Institute Foundation (2021). The authors are supported by the MIUR Progetto di Eccellenza.

Institutional Review Board Statement: Not applicable.

Informed Consent Statement: Not applicable.

Data Availability Statement: Not applicable.

Conflicts of Interest: The authors declare no conflict of interest relevant to the topic of the paper.

References

1. Ros, E.; Singh, A.; O'Keefe, J.H. Nuts: Natural Pleiotropic Nutraceuticals. *Nutrients* **2021**, *13*, 3269. [CrossRef] [PubMed]
2. Petrella, C.; Di Certo, M.G.; Gabanella, F.; Barbato, C.; Ceci, F.M.; Greco, A.; Ralli, M.; Polimeni, A.; Angeloni, A.; Severini, C.; et al. Mediterranean Diet, Brain and Muscle: Olive Polyphenols and Resveratrol Protection in Neurodegenerative and Neuromuscular Disorders. *Curr. Med. Chem.* **2021**, *28*, 7595–7613. [CrossRef] [PubMed]
3. Schwingshackl, L.; Morze, J.; Hoffmann, G. Mediterranean diet and health status: Active ingredients and pharmacological mechanisms. *Br. J. Pharmacol.* **2020**, *177*, 1241–1257. [CrossRef] [PubMed]
4. Willett, W.C.; Sacks, F.; Trichopoulou, A.; Drescher, G.; Ferro-Luzzi, A.; Helsing, E.; Trichopoulos, D. Mediterranean diet pyramid: A cultural model for healthy eating. *Am. J. Clin. Nutr.* **1995**, *61*, 1402S–1406S. [CrossRef] [PubMed]
5. Nestle, M. Mediterranean diets: Historical and research overview. *Am. J. Clin. Nutr.* **1995**, *61*, 1313S–1320S. [CrossRef]
6. Haber, B. The Mediterranean diet: A view from history. *Am. J. Clin. Nutr.* **1997**, *66*, 1053S–1057S. [CrossRef]
7. Bonaccio, M.; Iacoviello, L.; Donati, M.B.; de Gaetano, G. The tenth anniversary as a UNESCO world cultural heritage: An unmissable opportunity to get back to the cultural roots of the Mediterranean diet. *Eur. J. Clin. Nutr.* **2022**, *76*, 179–183. [CrossRef]
8. Scali, J.; Richard, A.; Gerber, M. Diet profiles in a population sample from Mediterranean southern France. *Public Health Nutr.* **2001**, *4*, 173–182. [CrossRef]
9. Magni, P.; Bier, D.M.; Pecorelli, S.; Agostoni, C.; Astrup, A.; Brighenti, F.; Cook, R.; Folco, E.; Fontana, L.; Gibson, R.A.; et al. Perspective: Improving Nutritional Guidelines for Sustainable Health Policies: Current Status and Perspectives. *Adv. Nutr.* **2017**, *8*, 532–545. [CrossRef]
10. Mithril, C.; Dragsted, L.O.; Meyer, C.; Blauert, E.; Holt, M.K.; Astrup, A. Guidelines for the New Nordic Diet. *Public Health Nutr.* **2012**, *15*, 1941–1947. [CrossRef]
11. U.S. Department of Agriculture; U.S. Department of Health and Human Services. ietary Guidelines for Americans, 2020–2025. Available online: https://dietaryguidelines.gov (accessed on 24 February 2022).
12. Hwalla, N.; Jaafar, Z.; Sawaya, S. Dietary Management of Type 2 Diabetes in the MENA Region: A Review of the Evidence. *Nutrients* **2021**, *13*, 1060. [CrossRef] [PubMed]
13. Doustmohammadian, A.; Clark, C.C.T.; Maadi, M.; Motamed, N.; Sobhrakhshankhah, E.; Ajdarkosh, H.; Mansourian, M.R.; Esfandyari, S.; Hanjani, N.A.; Nikkhoo, M.; et al. Favorable association between Mediterranean diet (MeD) and DASH with NAFLD among Iranian adults of the Amol Cohort Study (AmolCS). *Sci. Rep.* **2022**, *12*, 2131. [CrossRef] [PubMed]
14. Matana, A.; Franić, I.; Radić Hozo, E.; Burger, A.; Boljat, P. Adherence to the Mediterranean Diet among Children and Youth in the Mediterranean Region in Croatia: A Comparative Study. *Nutrients* **2022**, *14*, 302. [CrossRef] [PubMed]
15. Rubini, A.; Vilaplana-Prieto, C.; Flor-Alemany, M.; Yeguas-Rosa, L.; Hernández-González, M.; Félix-García, F.J.; Félix-Redondo, F.J.; Fernández-Bergés, D. Assessment of the Cost of the Mediterranean Diet in a Low-Income Region: Adherence and Relationship with Available Incomes. *BMC Public Health* **2022**, *22*, 58. [CrossRef]
16. Papadimitriou, A.; Foscolou, A.; Itsiopoulos, C.; Thodis, A.; Kouris-Blazos, A.; Brazionis, L.; Sidossis, A.C.; Polychronopoulos, E.; Kokkinos, P.; Panagiotakos, D.; et al. Adherence to the Mediterranean Diet and Successful aging in Greeks living in Greece and abroad: The epidemiological Mediterranean Islands Study (MEDIS). *Nutr. Health* **2022**, 2601060211072363. [CrossRef]
17. Šarac, J.; Havaš Auguštin, D.; Lovrić, M.; Stryeck, S.; Šunić, I.; Novokmet, N.; Missoni, S. A Generation Shift in Mediterranean Diet Adherence and Its Association with Biological Markers and Health in Dalmatia, Croatia. *Nutrients* **2021**, *13*, 4564. [CrossRef]
18. Obeid, C.A.; Gubbels, J.S.; Jaalouk, D.; Kremers, S.P.J.; Oenema, A. Adherence to the Mediterranean diet among adults in Mediterranean countries: A systematic literature review. *Eur. J. Nutr.* **2022**, 1–18. [CrossRef]
19. Magni, P.; Dozio, E.; Ruscica, M.; Celotti, F.; Masini, M.A.; Prato, P.; Broccoli, M.; Mambro, A.; Morè, M.; Strollo, F. Feeding behavior in mammals including humans. *Ann. N. Y. Acad Sci.* **2009**, *1163*, 221–232. [CrossRef]

20. Archero, F.; Ricotti, R.; Solito, A.; Carrera, D.; Civello, F.; Di Bella, R.; Bellone, S.; Prodam, F. Adherence to the Mediterranean Diet among School Children and Adolescents Living in Northern Italy and Unhealthy Food Behaviors Associated to Overweight. *Nutrients* **2018**, *10*, 1322. [CrossRef]
21. Chacón-Cuberos, R.; Badicu, G.; Zurita-Ortega, F.; Castro-Sánchez, M. Mediterranean Diet and Motivation in Sport: A Comparative Study Between University Students from Spain and Romania. *Nutrients* **2018**, *11*, 30. [CrossRef]
22. Tsakiraki, M.; Grammatikopoulou, M.G.; Stylianou, C.; Tsigga, M. Nutrition transition and health status of Cretan women: Evidence from two generations. *Public Health Nutr.* **2011**, *14*, 793–800. [CrossRef] [PubMed]
23. La Fauci, V.; Alessi, V.; Assefa, D.Z.; Lo Giudice, D.; Calimeri, S.; Ceccio, C.; Antonuccio, G.M.; Genovese, C.; Squeri, R. Mediterranean diet: Knowledge and adherence in Italian young people. *Clin. Ter.* **2020**, *171*, e437–e443. [CrossRef] [PubMed]
24. Sofi, F.; Cesari, F.; Abbate, R.; Gensini, G.F.; Casini, A. Adherence to Mediterranean diet and health status: Meta-analysis. *BMJ* **2008**, *337*, a1344. [CrossRef] [PubMed]
25. Guasch-Ferré, M.; Willett, W.C. The Mediterranean diet and health: A comprehensive overview. *J. Intern. Med.* **2021**, *290*, 549–566. [CrossRef]
26. de Lorgeril, M.; Salen, P.; Martin, J.L.; Monjaud, I.; Delaye, J.; Mamelle, N. Mediterranean diet, traditional risk factors, and the rate of cardiovascular complications after myocardial infarction: Final report of the Lyon Diet Heart Study. *Circulation* **1999**, *99*, 779–785. [CrossRef]
27. Estruch, R.; Ros, E.; Salas-Salvadó, J.; Covas, M.I.; Corella, D.; Arós, F.; Gómez-Gracia, E.; Ruiz-Gutiérrez, V.; Fiol, M.; Lapetra, J.; et al. Primary Prevention of Cardiovascular Disease with a Mediterranean Diet Supplemented with Extra-Virgin Olive Oil or Nuts. *N. Engl. J. Med.* **2018**, *378*, e34. [CrossRef]
28. Keys, A. Coronary heart disease, serum cholesterol, and the diet. *Acta Med. Scand.* **1980**, *207*, 153–160. [CrossRef]
29. Keys, A. Wine, garlic, and CHD in seven countries. *Lancet* **1980**, *315*, 145–146. [CrossRef]
30. Shikany, J.M.; Safford, M.M.; Soroka, O.; Brown, T.M.; Newby, P.K.; Durant, R.W.; Judd, S.E. Mediterranean Diet Score, Dietary Patterns, and Risk of Sudden Cardiac Death in the REGARDS Study. *J. Am. Heart Assoc.* **2021**, *10*, e019158. [CrossRef]
31. Wan, D.; Li, V.; Banfield, L.; Azab, S.; de Souza, R.J.; Anand, S.S. Diet and Nutrition in Peripheral Artery Disease: A Systematic Review. *Can. J. Cardiol.* **2022**, *38*, 672–680. [CrossRef]
32. Schwingshackl, L.; Missbach, B.; König, J.; Hoffmann, G. Adherence to a Mediterranean diet and risk of diabetes: A systematic review and meta-analysis. *Public Health Nutr.* **2015**, *18*, 1292–1299. [CrossRef] [PubMed]
33. Kastorini, C.M.; Milionis, H.J.; Esposito, K.; Giugliano, D.; Goudevenos, J.A.; Panagiotakos, D.B. The effect of Mediterranean diet on metabolic syndrome and its components: A meta-analysis of 50 studies and 534,906 individuals. *J. Am. Coll. Cardiol.* **2011**, *57*, 1299–1313. [CrossRef] [PubMed]
34. Martínez-González, M.A.; de la Fuente-Arrillaga, C.; Nunez-Cordoba, J.M.; Basterra-Gortari, F.J.; Beunza, J.J.; Vazquez, Z.; Benito, S.; Tortosa, A.; Bes-Rastrollo, M. Adherence to Mediterranean diet and risk of developing diabetes: Prospective cohort study. *BMJ* **2008**, *336*, 1348–1351. [CrossRef] [PubMed]
35. Mirabelli, M.; Chiefari, E.; Arcidiacono, B.; Corigliano, D.M.; Brunetti, F.S.; Maggisano, V.; Russo, D.; Foti, D.P.; Brunetti, A. Mediterranean Diet Nutrients to Turn the Tide against Insulin Resistance and Related Diseases. *Nutrients* **2020**, *12*, 1066. [CrossRef] [PubMed]
36. Bradbury, K.E.; Appleby, P.N.; Key, T.J. Fruit, vegetable, and fiber intake in relation to cancer risk: Findings from the European Prospective Investigation into Cancer and Nutrition (EPIC). *Am. J. Clin. Nutr.* **2014**, *100* (Suppl. S1), 394S–398S. [CrossRef]
37. Augimeri, G.; Bonofiglio, D. The Mediterranean Diet as a Source of Natural Compounds: Does It Represent a Protective Choice against Cancer? *Pharmaceuticals* **2021**, *14*, 920. [CrossRef]
38. Panza, F.; Solfrizzi, V.; Colacicco, A.M.; D'Introno, A.; Capurso, C.; Torres, F.; Del Parigi, A.; Capurso, S.; Capurso, A. Mediterranean diet and cognitive decline. *Public Health Nutr.* **2004**, *7*, 959–963. [CrossRef]
39. Solch, R.J.; Aigbogun, J.O.; Voyiadjis, A.G.; Talkington, G.M.; Darensbourg, R.M.; O'Connell, S.; Pickett, K.M.; Perez, S.R.; Maraganore, D.M. Mediterranean diet adherence, gut microbiota, and Alzheimer's or Parkinson's disease risk: A systematic review. *J. Neurol. Sci.* **2022**, *434*, 120166. [CrossRef]
40. Valls-Pedret, C.; Lamuela-Raventós, R.M.; Medina-Remón, A.; Quintana, M.; Corella, D.; Pintó, X.; Martínez-González, M.; Estruch, R.; Ros, E. Polyphenol-rich foods in the Mediterranean diet are associated with better cognitive function in elderly subjects at high cardiovascular risk. *J. Alzheimer's Dis.* **2012**, *29*, 773–782. [CrossRef]
41. Vlachos, G.S.; Yannakoulia, M.; Anastasiou, C.A.; Kosmidis, M.H.; Dardiotis, E.; Hadjigeorgiou, G.; Charisis, S.; Sakka, P.; Stefanis, L.; Scarmeas, N. The role of Mediterranean diet in the course of subjective cognitive decline in the elderly population of Greece: Results from a prospective cohort study. *Br. J. Nutr.* **2021**, 1–11. [CrossRef]
42. Coelho-Júnior, H.J.; Trichopoulou, A.; Panza, F. Cross-sectional and longitudinal associations between adherence to Mediterranean diet with physical performance and cognitive function in older adults: A systematic review and meta-analysis. *Ageing Res. Rev.* **2021**, *70*, 101395. [CrossRef] [PubMed]
43. Bonaccio, M.; Iacoviello, L.; de Gaetano, G.; Moli-Sani Investigators. The Mediterranean diet: The reasons for a success. *Thromb. Res.* **2012**, *129*, 401–404. [CrossRef] [PubMed]
44. Olmastroni, E.; Shlyakhto, E.V.; Konradi, A.O.; Rotar, O.P.; Alieva, A.S.; Boyarinova, M.A.; Baragetti, A.; Grigore, L.; Pellegatta, F.; Tragni, E.; et al. Epidemiology of cardiovascular risk factors in two population-based studies. *Atheroscler. Suppl.* **2018**, *35*, e14–e20. [CrossRef] [PubMed]

45. Olmastroni, E.; Baragetti, A.; Casula, M.; Grigore, L.; Pellegatta, F.; Pirillo, A.; Tragni, E.; Catapano, A.L. Multilevel Models to Estimate Carotid Intima-Media Thickness Curves for Individual Cardiovascular Risk Evaluation. *Stroke* **2019**, *50*, 1758–1765. [CrossRef] [PubMed]
46. Magni, P.; Baragetti, A.; Casula, M.; Grigore, L.; Olmastroni, E.; Pellegatta, F.; Catapano, A.L. Life-Style And Cardio-Metabolic Profile Of A Population Living In A Clustered Alpine Village (The Plic Chiesa Study): Comparison With An Urban Cohort (Plic Study). *Atherosclerosis* **2019**, *287*, e83–e84. [CrossRef]
47. Abellán Alemán, J.; Zafrilla Rentero, M.P.; Montoro-García, S.; Mulero, J.; Pérez Garrido, A.; Leal, M.; Guerrero, L.; Ramos, E.; Ruilope, L.M. Adherence to the "Mediterranean Diet" in Spain and Its Relationship with Cardiovascular Risk (DIMERICA Study). *Nutrients* **2016**, *8*, 680. [CrossRef]
48. Tsiampalis, T.; Faka, A.; Kouvari, M.; Psaltopoulou, T.; Pitsavos, C.; Chalkias, C.; Panagiotakos, D.B. The impact of socioeconomic and environmental determinants on Mediterranean diet adherence: A municipal-level spatial analysis in Athens metropolitan area, Greece. *Int. J. Food Sci. Nutr.* **2021**, *72*, 259–270. [CrossRef]
49. Kapelios, C.J.; Kyriazis, I.; Ioannidis, I.; Dimosthenopoulos, C.; Hatziagelaki, E.; Liatis, S.; Group, P.S. Diet, life-style and cardiovascular morbidity in the rural, free living population of Elafonisos island. *BMC Public Health* **2017**, *17*, 147. [CrossRef]
50. Kyprianidou, M.; Panagiotakos, D.; Faka, A.; Kambanaros, M.; Makris, K.C.; Christophi, C.A. Adherence to the Mediterranean diet in Cyprus and its relationship to multi-morbidity: An epidemiological study. *Public Health Nutr.* **2021**, *24*, 4546–4555. [CrossRef]
51. Sumalla-Cano, S.; Forbes-Hernández, T.; Aparicio-Obregón, S.; Crespo, J.; Eléxpuru-Zabaleta, M.; Gracia-Villar, M.; Giampieri, F.; Elío, I. Changes in the Lifestyle of the Spanish University Population during Confinement for COVID-19. *Int. J. Environ. Res. Public Health* **2022**, *19*, 2210. [CrossRef]
52. Di Renzo, L.; Gualtieri, P.; Pivari, F.; Soldati, L.; Attinà, A.; Cinelli, G.; Leggeri, C.; Caparello, G.; Barrea, L.; Scerbo, F.; et al. Eating habits and lifestyle changes during COVID-19 lockdown: An Italian survey. *J. Transl. Med.* **2020**, *18*, 229. [CrossRef] [PubMed]
53. Kouris-Blazos, A.; Gnardellis, C.; Wahlqvist, M.L.; Trichopoulos, D.; Lukito, W.; Trichopoulou, A. Are the advantages of the Mediterranean diet transferable to other populations? A cohort study in Melbourne, Australia. *Br. J. Nutr.* **1999**, *82*, 57–61. [CrossRef] [PubMed]
54. Chen, M.; Creger, T.; Howard, V.; Judd, S.E.; Harrington, K.F.; Fontaine, K.R. Geospatial analysis of Mediterranean diet adherence in the United States. *Public Health Nutr.* **2021**, *24*, 2920–2928. [CrossRef] [PubMed]
55. Knight, C.J.; Jackson, O.; Rahman, I.; Burnett, D.O.; Frugé, A.D.; Greene, M.W. The Mediterranean Diet in the Stroke Belt: A Cross-Sectional Study on Adherence and Perceived Knowledge, Barriers, and Benefits. *Nutrients* **2019**, *11*, 1847. [CrossRef]
56. Martínez-González, M.A.; García-Arellano, A.; Toledo, E.; Salas-Salvadó, J.; Buil-Cosiales, P.; Corella, D.; Covas, M.I.; Schröder, H.; Arós, F.; Gómez-Gracia, E.; et al. A 14-item Mediterranean diet assessment tool and obesity indexes among high-risk subjects: The PREDIMED trial. *PLoS ONE* **2012**, *7*, e43134. [CrossRef] [PubMed]
57. Panagiotakos, D.B.; Pitsavos, C.; Stefanadis, C. Dietary patterns: A Mediterranean diet score and its relation to clinical and biological markers of cardiovascular disease risk. *Nutr. Metab. Cardiovasc. Dis.* **2006**, *16*, 559–568. [CrossRef]
58. Serra-Majem, L.; Ribas, L.; Ngo, J.; Ortega, R.M.; García, A.; Pérez-Rodrigo, C.; Aranceta, J. Food, youth and the Mediterranean diet in Spain. Development of KIDMED, Mediterranean Diet Quality Index in children and adolescents. *Public Health Nutr.* **2004**, *7*, 931–935. [CrossRef]
59. Trichopoulou, A.; Kouris-Blazos, A.; Wahlqvist, M.L.; Gnardellis, C.; Lagiou, P.; Polychronopoulos, E.; Vassilakou, T.; Lipworth, L.; Trichopoulos, D. Diet and overall survival in elderly people. *BMJ* **1995**, *311*, 1457–1460. [CrossRef]
60. Hu, F.B.; Bronner, L.; Willett, W.C.; Stampfer, M.J.; Rexrode, K.M.; Albert, C.M.; Hunter, D.; Manson, J.E. Fish and omega-3 fatty acid intake and risk of coronary heart disease in women. *JAMA* **2002**, *287*, 1815–1821. [CrossRef]
61. Monteagudo, C.; Mariscal-Arcas, M.; Rivas, A.; Lorenzo-Tovar, M.L.; Tur, J.A.; Olea-Serrano, F. Proposal of a Mediterranean Diet Serving Score. *PLoS ONE* **2015**, *10*, e0128594. [CrossRef]
62. Dominguez, L.J.; Di Bella, G.; Veronese, N.; Barbagallo, M. Impact of Mediterranean Diet on Chronic Non-Communicable Diseases and Longevity. *Nutrients* **2021**, *13*, 2028. [CrossRef] [PubMed]
63. Bhattacharya, R.; Zekavat, S.M.; Uddin, M.M.; Pirruccello, J.; Niroula, A.; Gibson, C.; Griffin, G.K.; Libby, P.; Ebert, B.L.; Bick, A.; et al. Association of Diet Quality With Prevalence of Clonal Hematopoiesis and Adverse Cardiovascular Events. *JAMA Cardiol.* **2021**, *6*, 1069–1077. [CrossRef] [PubMed]
64. Mithril, C.; Dragsted, L.O. Safety evaluation of some wild plants in the New Nordic Diet. *Food Chem. Toxicol.* **2012**, *50*, 4461–4467. [CrossRef] [PubMed]
65. Mithril, C.; Dragsted, L.O.; Meyer, C.; Tetens, I.; Biltoft-Jensen, A.; Astrup, A. Dietary composition and nutrient content of the New Nordic Diet. *Public Health Nutr.* **2013**, *16*, 777–785. [CrossRef] [PubMed]
66. Saxe, H. The New Nordic Diet is an effective tool in environmental protection: It reduces the associated socioeconomic cost of diets. *Am. J. Clin. Nutr.* **2014**, *99*, 1117–1125. [CrossRef]
67. Galbete, C.; Kröger, J.; Jannasch, F.; Iqbal, K.; Schwingshackl, L.; Schwedhelm, C.; Weikert, C.; Boeing, H.; Schulze, M.B. Nordic diet, Mediterranean diet, and the risk of chronic diseases: The EPIC-Potsdam study. *BMC Med.* **2018**, *16*, 99. [CrossRef]

68. Klonizakis, M.; Bugg, A.; Hunt, B.; Theodoridis, X.; Bogdanos, D.P.; Grammatikopoulou, M.G. Assessing the Physiological Effects of Traditional Regional Diets Targeting the Prevention of Cardiovascular Disease: A Systematic Review of Randomized Controlled Trials Implementing Mediterranean, New Nordic, Japanese, Atlantic, Persian and Mexican Dietary Interventions. *Nutrients* **2021**, *13*, 3034. [CrossRef]
69. Krznarić, Ž.; Karas, I.; Ljubas Kelečić, D.; Vranešić Bender, D. The Mediterranean and Nordic Diet: A Review of Differences and Similarities of Two Sustainable, Health-Promoting Dietary Patterns. *Front. Nutr.* **2021**, *8*, 683678. [CrossRef]
70. Tsartsali, P.K.; Thompson, J.L.; Jago, R. Increased knowledge predicts greater adherence to the Mediterranean diet in Greek adolescents. *Public Health Nutr.* **2009**, *12*, 208–213. [CrossRef]
71. Noale, M.; Nardi, M.; Limongi, F.; Siviero, P.; Caregaro, L.; Crepaldi, G.; Maggi, S.; Group, M.D.F.S. Adolescents in southern regions of Italy adhere to the Mediterranean diet more than those in the northern regions. *Nutr. Res.* **2014**, *34*, 771–779. [CrossRef]
72. Morelli, C.; Avolio, E.; Galluccio, A.; Caparello, G.; Manes, E.; Ferraro, S.; De Rose, D.; Santoro, M.; Barone, I.; Catalano, S.; et al. Impact of Vigorous-Intensity Physical Activity on Body Composition Parameters, Lipid Profile Markers, and Irisin Levels in Adolescents: A Cross-Sectional Study. *Nutrients* **2020**, *12*, 741. [CrossRef] [PubMed]
73. Grosso, G.; Marventano, S.; Buscemi, S.; Scuderi, A.; Matalone, M.; Platania, A.; Giorgianni, G.; Rametta, S.; Nolfo, F.; Galvano, F.; et al. Factors associated with adherence to the Mediterranean diet among adolescents living in Sicily, Southern Italy. *Nutrients* **2013**, *5*, 4908–4923. [CrossRef] [PubMed]
74. Augimeri, G.; Galluccio, A.; Caparello, G.; Avolio, E.; La Russa, D.; De Rose, D.; Morelli, C.; Barone, I.; Catalano, S.; Andò, S.; et al. Potential Antioxidant and Anti-Inflammatory Properties of Serum from Healthy Adolescents with Optimal Mediterranean Diet Adherence: Findings from DIMENU Cross-Sectional Study. *Antioxidants* **2021**, *10*, 1172. [CrossRef] [PubMed]
75. Morelli, C.; Avolio, E.; Galluccio, A.; Caparello, G.; Manes, E.; Ferraro, S.; Caruso, A.; De Rose, D.; Barone, I.; Adornetto, C.; et al. Nutrition Education Program and Physical Activity Improve the Adherence to the Mediterranean Diet: Impact on Inflammatory Biomarker Levels in Healthy Adolescents From the DIMENU Longitudinal Study. *Front. Nutr.* **2021**, *8*, 685247. [CrossRef]
76. Caparello, G.; Galluccio, A.; Giordano, C.; Lofaro, D.; Barone, I.; Morelli, C.; Sisci, D.; Catalano, S.; Andò, S.; Bonofiglio, D. Adherence to the Mediterranean diet pattern among university staff: A cross-sectional web-based epidemiological study in Southern Italy. *Int. J. Food Sci. Nutr.* **2020**, *71*, 581–592. [CrossRef]
77. Ceraudo, F.; Caparello, G.; Galluccio, A.; Avolio, E.; Augimeri, G.; De Rose, D.; Vivacqua, A.; Morelli, C.; Barone, I.; Catalano, S.; et al. Impact of Mediterranean Diet Food Choices and Physical Activity on Serum Metabolic Profile in Healthy Adolescents: Findings from the DIMENU Project. *Nutrients* **2022**, *14*, 881. [CrossRef]
78. Hoffman, R.; Gerber, M. Evaluating and adapting the Mediterranean diet for non-Mediterranean populations: A critical appraisal. *Nutr. Rev.* **2013**, *71*, 573–584. [CrossRef]
79. Jin, Q.; Black, A.; Kales, S.N.; Vattem, D.; Ruiz-Canela, M.; Sotos-Prieto, M. Metabolomics and Microbiomes as Potential Tools to Evaluate the Effects of the Mediterranean Diet. *Nutrients* **2019**, *11*, 207. [CrossRef]
80. Mitsou, E.K.; Kakali, A.; Antonopoulou, S.; Mountzouris, K.C.; Yannakoulia, M.; Panagiotakos, D.B.; Kyriacou, A. Adherence to the Mediterranean diet is associated with the gut microbiota pattern and gastrointestinal characteristics in an adult population. *Br. J. Nutr.* **2017**, *117*, 1645–1655. [CrossRef]

MDPI
St. Alban-Anlage 66
4052 Basel
Switzerland
Tel. +41 61 683 77 34
Fax +41 61 302 89 18
www.mdpi.com

Nutrients Editorial Office
E-mail: nutrients@mdpi.com
www.mdpi.com/journal/nutrients

www.ingramcontent.com/pod-product-compliance
Lightning Source LLC
LaVergne TN
LVHW070842170726
843515LV00005B/555

* 9 7 8 3 0 3 6 5 5 0 9 5 4 *